# Composting in the Framework of Circular Economy

# Composting in the Framework of Circular Economy

Editor

**Antoni Sánchez**

MDPI

Basel • Beijing • Wuhan • Barcelona • Belgrade • Novi Sad • Cluj • Manchester

*Editor*
Antoni Sánchez
Department of Chemical,
Biological and Environmental
Engineering Autonomous
University of Barcelona
Bellaterra-Barcelona
Spain

*Editorial Office*
MDPI
St. Alban-Anlage 66
4052 Basel, Switzerland

This is a reprint of articles from the Special Issue published online in the open access journal *Processes* (ISSN 2227-9717) (available at: https://www.mdpi.com/journal/processes/special_issues/composting_framework_circular_economy).

For citation purposes, cite each article independently as indicated on the article page online and as indicated below:

| Lastname, A.A.; Lastname, B.B. Article Title. *Journal Name* **Year**, *Volume Number*, Page Range. |
|---|

**ISBN 978-3-7258-0165-7 (Hbk)**
**ISBN 978-3-7258-0166-4 (PDF)**
**doi.org/10.3390/books978-3-7258-0166-4**

# Contents

# About the Editor

**Antoni Sánchez**

Antoni Sánchez has been leading the Composting Research Group since 2000. He works as a Full Professor in the Department of Chemical, Biological and Environmental Engineering of Universitat Autònoma de Barcelona. His research lines cover the main aspects involved in the treatment of organic solid waste, through several strategies. Composting, anaerobic digestion and solid-state fermentation are the main lines. He has participated and coordinated several Spanish and European projects on this topic, supervised PhD and Master Thesis and he is the author of more than 200 indexed papers as well as book chapters and patents.

processes

MDPI

*Editorial*

# Special Issue on "Composting in the Framework of a Circular Economy"

**Antoni Sánchez**

Department of Chemical, Biological and Environmental Engineering, Autonomous University of Barcelona, Bellaterra, 08193 Barcelona, Spain; antoni.sanchez@uab.cat

Composting has been recognized as a sustainable technology to treat and manage organic waste. Based on the principles of composting published in Haug's classical handbook thirty years ago [1], compost research has been regularly published in the environmental scientific literature, ranging from the first typical composting works, where the evolution of some parameters were monitored, to current studies focused on advanced aspects of composting, such as the microbiology of composting and its role in complex biorefineries. In general, composting is a technology that, together with its closer relative solid-state fermentation, is destined to change the paradigm: from waste to resources.

In parallel with this increase in research activity and knowledge, composting has gained the acceptance of society. Once its main initial problems, typically related to odors, have been overcome [2], composting will be massively implemented worldwide. Three factors will play a key role in this expansion:

(1) Enforcing new regulations that help avoid or tax the disposal of organic matter in sanitary landfills and convert the source selection of domestic organic matter through mandatory rules. This is the case in Europe, for example.
(2) Desertification in many parts of the world, which implies a lack of organic matter; compost is well known for its restoration purposes.
(3) The flexibility of composting, which makes this technology adaptable to a wide range of organic waste, climate conditions, sizes, and dimensions, ranging from simple turned-pile systems to highly controlled bioreactors.

In summary, composting is experiencing a "second youth" and the number of plants, as well as the uses of compost, is increasing exponentially all over the world. Recently, the same phenomenon can be observed with the "sister" technology of composting: anaerobic digestion. Today, the exponential growth of anaerobic digestion is very perceptible in some parts of the world because of the need for renewable and locally available energy. Although anaerobic digestion is not as flexible and robust as composting, it is a technology with a negative carbon balance in terms of global warming, which makes it very attractive [3]. Additionally, anaerobic digestion results in digestate, making it a very good substrate for composting.

The papers published in this Special Issue on "Composting in the Framework of a Circular Economy" are good examples of the interest in composting. Although composting has been widely studied and thousands of excellent papers can be found in the scientific literature, it is also a very complex process that has critical implications in fields such as solid-state heat and mass transfer, multiphasic reactions and bioreactors, microbiology, and scale-up processes, among others [4]. Therefore, any composting researcher agrees that composting and compost research is far from being finished [5].

In this Special Issue, we collected relevant articles related to the current aspects of composting research. It is also worthwhile to mention that some of the articles are from developing countries, especially in South America and Africa, where composting is in the initial stage of being a substitute to traditional landfills, which have severe sanitary and environmental problems.

**Citation:** Sánchez, A. Special Issue on "Composting in the Framework of a Circular Economy". *Processes* **2023**, *11*, 1573. https://doi.org/10.3390/pr11051573

Received: 10 April 2023
Accepted: 19 May 2023
Published: 21 May 2023

Regarding some of the topics presented in this Special Issue, one review paper is related to the gaseous emissions from composting [6]. This topic is critical for the proper development of composting plants, since it determines the environmental impact of composting, as well as possible influences on the stakeholders that finally decide to use composting for managing organic waste. The review provides the updated state of the art of these emissions, together with good practices and abatement technologies to mitigate the effect of these gases. The addition of biochar, for instance, to remove the emissions of methane and nitrous oxide (two powerful greenhouse gases) and volatile organic compounds causing unpleasant odors is in an early stage of research.

Another important research line treated in this Special Issue is the agronomic value of composting. This is critical for the restoration of polluted soils, or simply for soils with low levels of organic matter. One article is dedicated to assessing the agronomic value of the material resulting from the compost-bedded pack in dairy barns, with this resulting material having potentially valuable characteristics such as organic amendment of the soil [7]. Another paper examines the regional management of hospitality food waste by exploiting the municipal waste management infrastructure and carrying out intensive composting at the source [8]. The results show that the co-maturation experiment with animal by-products and municipal green waste primary composts proves that the phytotoxicity parameters of the cured compost were in the optimal range or below the thresholds. Both papers are good examples on how diverse materials can result in compost with evident benefits to soil. A very interesting paper is also presented that addresses the effect of compost in favoring the growth of specific vegetal clones, and how some genotypes are sensitive to the use of different substrates, which is a scarcely treated topic. These clones might be a good option for evaluating compost-based substrates for forestry applications [9]. Finally, a review is presented on the fate of the major components of organic matter, such as C, N, P, and K, during the composting process, a point that is critical for exploiting the benefits of compost in soil [10].

The rest of the papers see composting researchers exploring new findings on the use of specific inoculums for composting, technology advances, or new analytical tools to monitor the process.

To summarize, this Special Issue compiles useful information for composting researchers, including reviews and original research papers which give the reader a reliable picture of what fields of composting research are predominant in the current literature.

**Conflicts of Interest:** The author declares no conflict of interest.

## References

1. Haug, R.T. *The Practical Handbook of Compost Engineering*; CRC Press: Boca Raton, FL, USA, 1993.
2. González, D.; Gabriel, D.; Sánchez, A. Odors Emitted from Biological Waste and Wastewater Treatment Plants: A Mini-Review. *Atmosphere* **2022**, *13*, 798. [CrossRef]
3. Colón, C.; Cadena, E.; Pognani, P.; Barrena, R.; Sánchez, A.; Font, X.; Artola, A. Determination of the energy and environmental burdens associated with the biological treatment of source-separated Municipal Solid Wastes. *Energy Environ. Sci.* **2012**, *5*, 5731–5741. [CrossRef]
4. Cerda, A.; Artola, A.; Font, X.; Barrena, R.; Gea, T.; Sánchez, A. Composting of food wastes: Status and challenges. *Bioresour. Technol.* **2018**, *248A*, 57–67. [CrossRef] [PubMed]
5. Komilis, D.P. Compost quality: Is research still needed to assess it or do we have enough knowledge? *Waste Manag.* **2015**, *38*, 1–2. [CrossRef] [PubMed]
6. Sayara, T.; Sánchez, A. Gaseous Emissions from the Composting Process: Controlling Parameters and Strategies of Mitigation. *Processes* **2021**, *9*, 1844. [CrossRef]
7. Llonch, L.; Gordo, C.; López, M.; Castillejos, L.; Ferret, A.; Balanyà, T. Agronomic Characteristics of the Compost-Bedded Pack Made with Forest Biomass or Sawdust. *Processes* **2021**, *9*, 546. [CrossRef]
8. Stunžėnas, E.; Kliopova, I.; Kliaugaitė, D.; Pranas Budrys, R. Industrial Symbiosis for Optimal Bio-Waste Management and Production of a Higher Value-Added Product. *Processes* **2021**, *9*, 2228. [CrossRef]

9. Yang, Z.; Muhayodin, F.; Larsen, O.C.; Miao, H.; Xue, B.; Rotter, V.S. A Review of Composting Process Models of Organic Solid Waste with a Focus on the Fates of C, N, P, and K. *Processes* **2021**, *9*, 473. [CrossRef]
10. Rizzo, P.F.; Salinas, M.C.; Della Torre, V.; Diez, J.P.; Sallesses, L.F.; Riera, N.I.; Pathauer, P.S.; Komilis, D.; Sánchez, A. Effect of Poultry Manure-Derived Compost on the Growth of *eucalypts* spp. Hybrid Clones. *Processes* **2022**, *10*, 2182. [CrossRef]

processes

MDPI

*Article*

# Incorporation of Substrates and Inoculums as Operational Strategies to Promote Lignocellulose Degradation in Composting of Green Waste—A Pilot-Scale Study

**Edgar Ricardo Oviedo-Ocaña [1,*], Jonathan Soto-Paz [1], Viviana Sanchez-Torres [2] and Antoni Sánchez [3]**

1 Escuela de Ingeniería Civil, Universidad Industrial de Santander, Bucaramanga 680002, Colombia
2 Escuela de Ingeniería Química, Universidad Industrial de Santander, Bucaramanga 680002, Colombia
3 Composting Research Group, Department of Chemical Engineering, Universitat Autònoma de Barcelona, 08193 Barcelona, Spain
* Correspondence: eroviedo@uis.edu.co

**Abstract:** Composting is a sustainable alternative for green waste (GW) valorization contributing to the circular bioeconomy. However, the processing time must be reduced and the end-product quality must be improved. This study determined the effect of the incorporation of processed food waste (PFW), unprocessed food (UPFW), sawdust (SW), phosphate rock (PR) and a specific bacterial inoculum on GW-composting process parameters and product quality. Three treatments were evaluated in 120 kg piles: (i) TA: (GW + UPFW + PFW + inoculum), (ii) TB (GW + UPFW + PFW), and (iii) TC (GW). An inoculum of *Bacillus* sp. and *Paenibacillus* sp. was incorporated in the cooling phase for TA. On the other hand, the effect of the inoculum at the laboratory scale (20 kg reactors) was compared with that found at the pilot scale (120 kg piles). The incorporation of FW, SW, PR and the inoculum increased the amount of lignocellulose biodegradation (TA: 29.1%; TB: 22.7%; TC: 18.2%), which allowed for a reduction of up to 14 days of processing time. The product obtained for TA had a similar quality to the other two treatments, although a lower phytotoxicity was determined according to the germination index (TA: 95%; TB: 85%; and TC: 83%). The final product of TA showed the best agricultural characteristics with pH 8.3, TOC of 24.8%, TN of 1.32%, and GI of 98.8%. Finally, the scaling effect with the bacterial inoculum was shown to affect parameters such as the TOC, TN, GI, and, to a lesser extent, temperature and pH. The results obtained in this paper highlight the importance of optimizing the composting of GW, specifically with the use of co-substrates and specific inocula, which can be of interest for composting materials with a high content of lignocellulose such as GW.

**Keywords:** green waste; composting; bacteria; food waste; lignocellulose

**Citation:** Oviedo-Ocaña, E.R.; Soto-Paz, J.; Sanchez-Torres, V.; Sánchez, A. Incorporation of Substrates and Inoculums as Operational Strategies to Promote Lignocellulose Degradation in Composting of Green Waste—A Pilot-Scale Study. *Processes* **2023**, *11*, 241. https://doi.org/10.3390/pr11010241

Academic Editor: Elsayed Elbeshbishy

Received: 30 November 2022
Revised: 19 December 2022
Accepted: 7 January 2023
Published: 11 January 2023



## 1. Introduction

The management of green waste (GW) is a challenge due to its heterogeneous composition (i.e., wood, branches, leaves, soil, and grass clippings) and predominance of lignocellulosic compounds such as cellulose (40%), hemicellulose (20–30%), and lignin (20–30%). Although composting can be used to transform GW into a product with potential agricultural value, the difficulty in degrading lignocellulose increases the processing time and reduces the product quality [1].

To optimize GW composting, various strategies have been developed to accelerate the biodegradability of lignocellulosic compounds [1,2]. The incorporation of substrates such as sawdust (SW) and phosphate rock (PR) have been used to provide porosity and phosphorus [3,4]; however, few studies have evaluated the amendment addition of food waste (FW) [5,6], which constitutes more than 70% of municipal solid waste in developing countries [7].

The incorporation of microbial inoculants during the cooling phase is a strategy that represents a key operational change to reduce processing time [8]. However, some studies have reported that there is no significant effect associated with the use of microbial inoculants due to the potential competition between the exogenous species and the indigenous microbial species of the process [9]. Therefore, further studies are needed to evaluate the synergistic or antagonistic effects of microbial inoculation on the GW-composting process and end-product quality [8]. Bacterial strains from the genus *Bacillus* have been used as inoculants for GW composting due to their capacity to degrade lignocellulose and the formation of humic substances [10]. In contrast, bacteria from the genus *Paenibacillus* have been little studied in the composting of GW with other substrates despite their potential to secrete specialized enzymes for lignocellulose degradation [11].

Additionally, the scaling up of the composting process is considered fundamental for the optimization of the process and the evaluation of its applicability on a large scale [8]. Previously, the effects of inoculation in the cooling phase with *Bacillus* sp. and *Paenibacillus* sp. were preliminarily evaluated at the laboratory scale (500 mL reactors) [12] and in reactors with a capacity of 20 kg for GW and FW co-composting amended with SW and PR [10]. In the present study, the effect of inoculation with *Bacillus* sp. and *Paenibacillus* sp. on GW and FW co-composting was scaled up to a pilot scale of 120 kg composting piles. This study contributes to the search for options to improve GW-composting implementation for urban GW management, which can be implemented at the full scale. Furthermore, this study provides elements for the recovery of waste to generate valuable products within the framework of the circular bioeconomy.

## 2. Materials and Methods

### 2.1. Substrates and Composting Process

GW was collected from the campus of Universidad Industrial de Santander (Colombia). FW was obtained from a marketplace where PFW and UPFW were separated at the source. SW and PR were purchased from a commercial establishment. SW provided carbon, porosity, and adjusted moisture content [13], while PR provided phosphorus and porosity [3]. The substrates were manually mixed in the following proportions: 50% GW, 32.5% UFW, 2.5% PFW, 13% SW, and 2% PR. The substrate mixture was defined from an experiment carried out at the laboratory scale in 0.5 L reactors where the substrate mixture and the concentration of the bacterial strains used as the inoculum were optimized [12]. In addition, the mixture guarantees a C:N ratio greater than 25, a value recommended for the composting of green waste [14]. The experiments were carried out at the pilot scale in conical piles of 120 kg. We believe that this study can be easily interpreted and its conclusions are useful for works at the full scale. The piles were set up leveled on concrete and fenced with polyethylene shade cloth to prevent access by external agents. In treatment A (TA), a mixture of substrates plus inoculum was studied; treatment B (TB) corresponded to a mixture of substrates (uninoculated); and in treatment C (TC), only GW was processed. Each treatment had a replicate. Before starting the experiment, the GW, UFW and PFW were manually crushed to a particle size of between 30 and 50 mm [1].

The physicochemical characteristics of the substrates and co-substrates were:

- PFW: moisture of 75.5 ± 7.6%, pH of 4.9 ± 0.4, EC of 3.1 ± 0.4 mS/cm, TOC of 33.5 ± 5.9% (db), TN of 1.2 ± 0.6% (db), and lignocellulose of 17.0 ± 2.6% (db).
- UFW: moisture of 79.1 ± 8.3%, pH of 5.1 ± 0.3, EC of 3.1 ± 0.4 mS/cm, TOC of 33.5 ± 5.9% (db), TN of 1.2 ± 0.6% (db) and lignocellulose of 17.8 ± 3.3% (db).
- TA and TB (including 50% GW, 32.5% UFW, 2.5% PFW, 13% SW, and 2% PR): moisture of 58.2 ± 2.5%, pH of 6.3 ± 0.2, EC of 3.5 ± 0.4 mS/cm, TOC of 47.7 ± 3.1% (db), TN of 1.7 ± 0.3% (db) and lignocellulose of 23.8 ± 1.9% (db).
- TC (100% GW): moisture of 27.3 ± 4.9%, pH of 6.9 ± 0.1, EC of 3.0 ± 0.3 mS/cm, TOC of 26.6 ± 5.8% (db), TN of 11.2 ± 0.5% (db) and lignocellulose of 35.1 ± 6.1% (db).

The bacterial inoculum consisted of *Bacillus* sp. and *Paenibacillus* sp. with concentrations of $4.85 \times 10^5$ CFU $mL^{-1}$ and $1.44 \times 10^5$ CFU $mL^{-1}$, respectively. The strains were individually cultivated in a Luria–Bertani medium at 37 °C and 200 rpm, sequentially scaling up in reactors with a capacity of 0.02, 0.06, 0.45, 1, and 7.5 L. The characteristics of the bioreactor for inoculum scaling are presented in more detail in the work of Oviedo et al. [10]. The inoculum was added to the piles at the start of the cooling phase (i.e., when the temperature was close to 45 °C).

Manual turns were applied depending on the temperature in the piles to maintain the aeration of the piles. When the temperature in the pile remained constant for three days, turning was applied. Furthermore, weekly humidification was applied to maintain a moisture content of between 40% and 60% [4] and stimulate biological activity [15]. The temperature was monitored daily using a digital thermometer with a resolution of 0.1 °C at five characteristic points of the reactors (i.e., centroid and four opposite points of the pile). Manual turnings were performed depending on the temperature reached. The monitoring was performed until all treatments reached ambient temperature.

On the other hand, data from a previous study by Oviedo et al. [10] that was carried out with the same treatments in 20 kg reactors were collected. The reactor configuration is presented in greater detail in the work of Rawoteea et al. [16]. With the information collected in this study on the laboratory scale, it was determined whether there was a scaling effect with the incorporation of the bacterial inoculum.

### 2.2. Analytical Methods

Five sub-samples of 300 g were obtained from the center and perimetral points of each pile at 30 cm of depth; the combined sample was used for laboratory analysis. The moisture content was gravimetrically measured at 70 °C in an oven until a constant weight was reached. The pH was determined using the potentiometric method (sensIONTM pH meter + MM374) in a suspension of 10 g of the processed composting material in 50 mL of distilled water (i.e., ratio of 1:10, *w*/*v*). Electrical conductivity (EC) was determined with the potentiometric method. The Total Kjeldahl Nitrogen (TN) was measured through titration according to Colombian Technical Standard NTC 5167. The ash content was determined with gravimetry at 550 °C, and the total organic carbon (TOC) was estimated from this value [17]. The organic matter losses were determined according to the work of Jiang et al. [18]. The stability of the product was determined with the Rottegrade self-heating test [19] once the process was finished. The germination index (GI) was determined using *Raphanus sativus* seeds, taking one gram of solid sample and diluting it with distilled water in a 1:10 (p/v) ratio [20]. As a maturity criterion, a GI of greater than 80% was adopted.

All experiments were carried out in triplicate; data were subjected to an analysis of variance (ANOVA), and significantly different means were evaluated using the least significant difference (LSD) at a significance level of $\alpha = 0.05$. Statistical analysis was performed in SPSS® Version 16.0 (Statistical Package for the Social Science, SPSS, Inc., Chicago, IL, USA).

### 2.3. Lignocellulose Degradation

The concentration of lignocellulose was determined according to the protocol of the National Renewable Energy Laboratory (NREL) with Soxhlet equipment (Model B-324, Buchi, Spain) considering the moisture, ash content at 550 °C and aqueous and organic extractives of the sample [21]. Additionally, the percentage of lignocellulose degradation was quantified once a week according to Equation (1).

$$\%Lignocellulose\ degradation = 1 - \frac{L_f}{L_i} * 100\% \quad (1)$$

where $L_i$ refers to the lignocellulose content at the beginning of the experiment and $L_f$ is the lignocellulose content at the time of measurement.

### 2.4. Product Quality

The end-product was manually sieved with a 1.25 cm mesh. An integrated sample of 4 kg was obtained for each treatment by combining sub-samples taken at the perimeter and centroid points of each pile. Quality parameters such as the pH, EC, TOC, TN, stability, and

GI were determined in triplicate. The results obtained in this study were compared with those obtained from a previous experiment with the same treatments but at the laboratory scale (20 kg reactors) [10].

## 3. Results

### 3.1. Physicochemical Changes during the Process

The temperature profiles of each treatment during the process are shown in Figure 1. The mesophilic phase in TA and TB was present until day 1 of the process, and it lasted until day 4 for TC. These results were similar to those previously indicated in the composting of UPFW, FW, and GW [22]. The longer time in TC was caused by the greater presence of slow-biodegrading organic carbon [1,23].

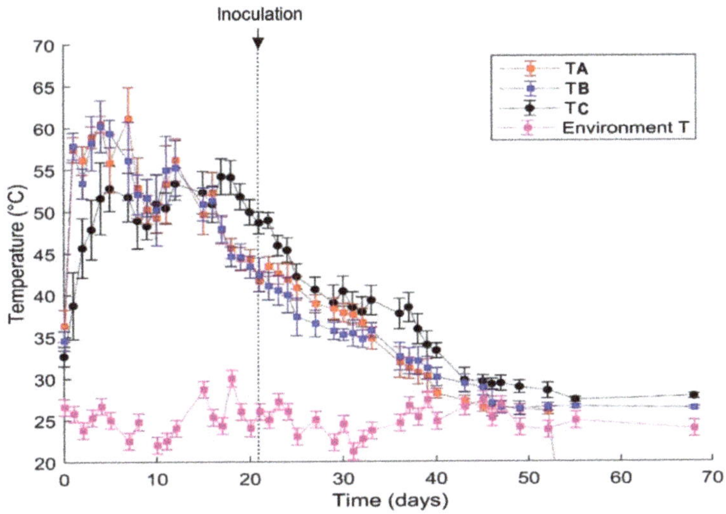


**Figure 1.** Temperature profiles during the composting process.

In the thermophilic phase, TA reached the highest temperature (64 °C) after 8 days of processing. In contrast, in TB, the highest temperature of 60.7 °C was reached at 9 days, showing no significant differences with TC. Regarding the sanitation of the process, if the temperature was maintained above 55 °C for three days, the material would be free of pathogens and weeds [22]. This condition was achieved with all treatments. The results indicated that the incorporation of FW, SW, and PR in TA and TB had a significant ($p < 0.05$) and synergistic effect on GW composting [13]. This resulted in an advantage compared with what has been documented in the composting of just GW due to the biological stimulus of microorganisms that results in the greater degradation of OM and therefore an increase in temperature. For TC, no temperature differences were observed with respect to what has been documented in the literature. This may have been associated with the predominance of lignocellulose, which is difficult to biologically degrade and therefore limits temperature increases. During the cooling phase, the incorporation of bacterial inocula can affect bacterial communities, as shown in another study [15]. The incorporation of the bacterial inoculum in TA increased the temperature (i.e., Δ 5 °C) between days 19 and 27 of the process. This evidenced that the inoculum stimulated the degradation of organic matter and lignocellulose, possibly due to the secretion of lignocellulolytic enzymes [24]. Similar results were reported with the inoculation with ammonifying bacteria, nitrobacteria, and *Azotobacter* in pig manure composting [24]. Then, the temperature dropped to 27 ± 3 °C on day 38 of the process. In contrast, TB reached this condition on day 43 and TC required 47 days. Therefore, the inoculum reduced the duration of the cooling phase by 13% and

23% compared with TB and TC, respectively. In addition, the duration of the cooling phase between TA and TC ($p = 0.0205$) showed significant differences.

The results found at the pilot scale showed trends similar to those found in the study of Oviedo et al. on 20 kg reactors [10]. In both cases, the inoculation increased the temperature gradients during the thermophilic phase (3–5 °C). However, the duration of both peaks differed between the scales (i.e., 10 and 6 days for the pilot and laboratory scale, respectively). These differences were probably caused by the mass of the material in the process and the heat-diffusion transport phenomena that prolonged the duration of the temperature peak at the pilot scale. In addition, the possible influence of exogenous microorganisms that could benefit from the incorporation of the inoculum is not ruled out.

Regarding the sanitation product, at both scales, the incorporation of the inoculum did not generate an antagonistic effect on the quality of the final product. Likewise, there was no significant reduction in the processing time with respect to the treatment that only included co-substrates ($p > 0.05$).

Regarding pH, TA and TB started with slightly acidic pH values (i.e., 5.8 and 6.3 units, respectively), as shown in Figure 2. During the thermophilic phase, the pH in all treatments increased to values higher than 8.0, probably due to ammonia volatilization as a consequence of the high temperatures reached [25]. Zhou et al. [26] reported similar pH profiles (7–9 units). In the cooling phase, there were no significant differences between treatments. At the end of the process, the lowest pH value was found in TA, which may have been associated with the nitrification process [26] and the synthesis of phenolic compounds [27]. A similar result was found in a study of the composting of straw residues, when an inoculum consisting of *Aeromonas caviae* sp., *Shinella* sp., *Rhizobium* sp., *Corynebacterium pseudotuberculosis* sp., and *Streptomyces clavuligerus* sp. was applied at the beginning of the cooling phase, which affected the pH dynamics [28]. In contrast, TC presented the highest pH value (i.e., 8.65) at the end of the process in this study.

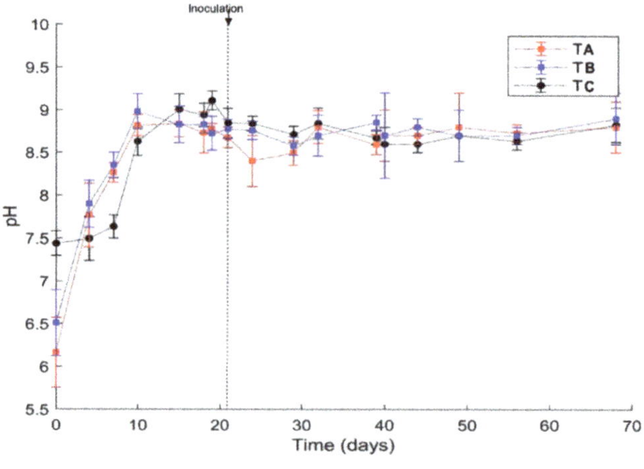


**Figure 2.** pH changes during the composting process.

The results found at the pilot scale were similar to those found at a smaller scale. The differences in terms of pH during the process steps were insignificant. Similarities were found in the rise of pH to alkaline values and a tendency to remain between 8 and 9 units in the cooling stage. These similarities were associated with the contribution of precedent proteins from PFW and UPFW and their mineralization during the ammonia volatilization due to the high temperatures reached [10]. In addition, the bacterial inoculum could affect the behavior of pH over time, as indicated in other investigations [8].

Figure 3 shows the dynamics of the TOC during composting. The concentration of the TOC in all treatments decreased over time. During the thermophilic stage in TA and TB, a higher degradation of the TOC (19 ± 3%) was achieved compared with TC (11 ± 1.2%). The differences between treatments was associated with the presence of FW, PR, and AS. Bohacz et al. [29] indicated that the incorporation of co-substrates stimulates biological activity, thus allowing for the higher degradation of the TOC and higher temperatures. After inoculation, a more pronounced decrease in the TOC was observed in TA and showed significant differences with respect to TB ($p = 0.038$). These results were associated with the incorporation of the inoculum that could stimulate the enzymatic activity and the mineralization of organic matter [30]. At the end of the process, the TOC losses for TA were 35.7%. This value was higher than those found in TB (32.1%) and TC (20.1%), indicating a synergistic effect of the inoculum on the process.

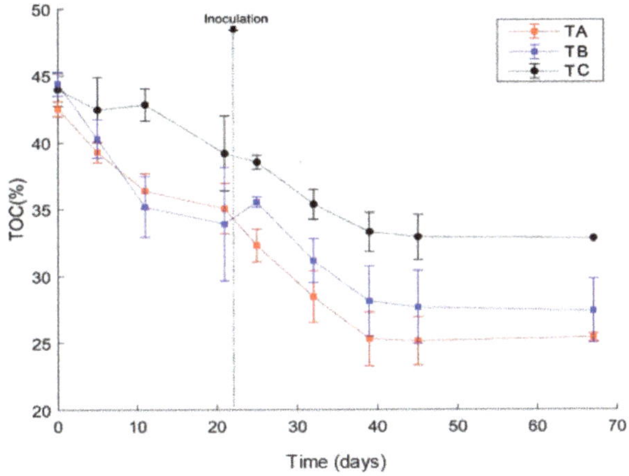


**Figure 3.** TOC changes during the composting process.

An effect of the inoculum was evidenced in the different evaluated scales. At the laboratory scale, there were greater TOC losses (41.1%) compared with what was found in the present pilot study (35.7%). These differences may be associated with the fact that the effect of the inoculum was enhanced in the more controlled environment and with a lower incidence of exogenous microorganisms due to the environmental and operational conditions of the laboratory-scale reactors. Additionally, oxygen diffusion was better at this scale due to the holes the presented reactors. This could have stimulated biological activity and promoted the degradation of OM. In contrast, at the pilot scale, the effect of compaction and clumping by the inoculum reduced the porosity of the material being processed and may have adversely affected mineralization.

On the other hand, the concentration of TN in all treatments increased over time (see Figure 4). The highest TN concentration was found in TC (2.4 ± 0.2%) and was associated with the availability of this element in the green waste retained by lignin [29]. This was associated with the fact the inoculum utilized the available nitrogen during the cooling phase for their metabolism and secretion of enzymes such as xylanase and cellulase [28,29]. According to Feng et al. [31], these enzymes are key to the degradation of lignocellulose. At the end of the process, the lowest concentration of 1.6 ± 0.3% was obtained in TA, and a concentration of 2.2 ± 0.2% was achieved in TB. The results indicated that the inoculum had an adverse effect on the availability of TN in the final product (i.e., possibly associated with a higher nutrient requirement due to higher biological activity).

The dynamics of nitrogen were similar between the considered scales. In both cases, the incorporation of the bacterial inoculum generated a decrease in the TN concentration, while in the other treatments (i.e., without the incorporation of the inoculum), the NT increased due to the effect of the OM concentration. The main difference was found in terms of NT concentration (i.e., 1.6 and 1.2% for the pilot and laboratory scales, respectively) during the cooling stage. In both cases, the effect of the bacterial inoculum was antagonistic to the potential to reduce the agricultural quality of the product.

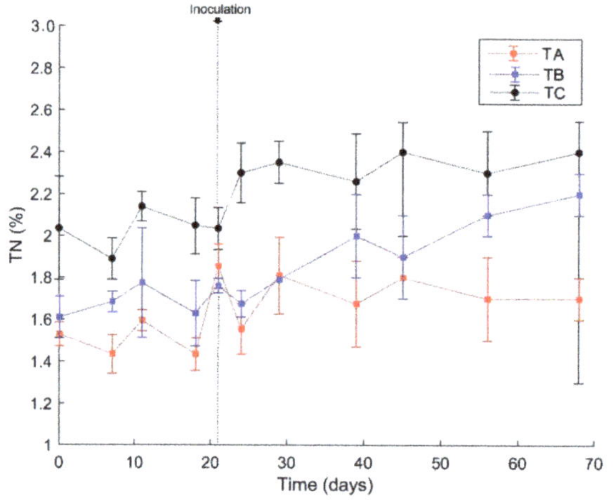


**Figure 4.** TN changes during the composting process.

### 3.2. *Biodegradation of Lignocellulose*

The percentage of lignocellulose in the composting material over time is shown in Figure 5. At the beginning of the process (i.e., mesophilic phase), the lignocellulose degradation was limited. This was mainly observed in TC, in which the predominance of organic compounds of difficult degradation could have limited the biological activity, thus prolonging the duration of the composting process. In contrast, in TA and TB, the biological activity was stimulated, and rapidly degrading carbon compounds were consumed, which increased the temperature and thus enhanced the hydrolysis of hemicellulose and cellulose when the temperature exceeded 60 °C [32].

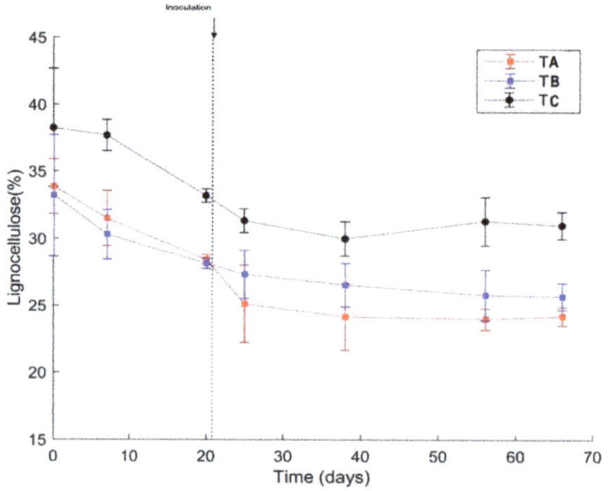


**Figure 5.** Percentage of lignocellulose in the composting material over time.

The lignocellulose degradation increased in all treatments after the start of the composting process. During the cooling phase, the percentage of lignocellulose degradation was 29.1%, 22.7%, and 18.2% for TA, TB, and TC, respectively. These results are consistent with those reported previously [8] due to the degradation of organic matter, the synthesis of phenolic compounds, and the synthesis of humic substances. The higher lignocellulose degradation for TA is associated with a higher content of TN that promoted the enzymatic activity of cellulose, xylanase, and phenoloxidase, among other lignocellulolytic enzymes [32,33]. Likewise, the pH values closest to neutrality in TA could have stimulated microbial activity [33]. Furthermore, it was previously reported that delignification takes place faster under neutral pH than under acidic or alkaline pH [34]. The lignin biodegradation observed in this study was higher than that reported previously (27.81%) for GW composting inoculated with a complex, non-defined microbial inoculum [8].

On the other hand, lignocellulose biodegradation was higher at the laboratory scale (31.7%) compared with that obtained at the pilot scale (29.1%). The differences could be associated with the fact that the environmental conditions at the laboratory scale favored the biological activity of the bacterial inoculum, promoting the degradation of lignocellulose. However, the effect of the bacterial inoculum showed a synergistic effect on both evaluated scales, evidencing higher efficiencies (28%) compared with that indicated in other works [8].

### 3.3. Product Quality

The physicochemical parameters of the obtained products are shown in Table 1. pH values in the alkaline range were obtained for the three treatments without significant differences between treatments ($p = 0.15$). The pH values in this study were similar to those reported (i.e., 7.5–9.0) in other lignocellulosic waste composting studies [35,36] This type of product has the potential to be used in soils with acidic characteristics [37]. Likewise, the products comply with the Colombian Technical Standard of Quality (NTC 5167).

**Table 1.** Physicochemical characteristics of the obtained compost product.

| Treatment | pH | EC | TOC | TN | GI |
|---|---|---|---|---|---|
| | | dS/m | %, db | %, db | % |
| TA | 8.4 ± 0.4 [a] | 1.5 ± 023 [a] | 25.4 ± 0.3 [a] | 1.7 ± 0.8 [a] | 95.8 ± 1.4 [b] |
| TB | 8.7 ± 0.3 [a] | 1.3 ± 0.3 [a] | 27. 4 ± 2.4 [a] | 2.2 ± 1.0 [b] | 85.4 ± 1.2 [c] |
| TC | 8.6 ± 0.1 [a] | 1.4 ± 0.2 [a] | 32.8 ± 1.7 [b] | 2.4 ± 1.1 [b] | 83.1 ± 2.1 [a] |

Note: EC: electrical conductivity; TOC: total organic carbon; TN: total nitrogen. TA: 50% GW, 32.5% UFW, 2.5% PFW, 13% SW, and 2% PR with bacterial inoculum. TB: 50% GW, 32.5% UFW, 2.5% PFW, 13% SW, and 2% PR (uninoculated). TC: 100% GW. The same letters indicate no significant differences ($p > 0.05$). Different letters indicate significant differences between treatments for the response parameter ($p < 0.05$).

The EC for the three treatments was in the range of 1.8–1.9 dS/m; because this was lower than 3 dS/m, the products are not considered phytotoxic for seeds [38]. Furthermore, no significant differences in EC were found between the products ($p = 0.17$) of the three treatments; therefore, the inoculation or the presence of the mixture of substrates did not affect this parameter. The EC values obtained in this study were lower than those previously reported in the co-composting of green waste and food waste inoculated with lactic acid bacteria, yeasts, and phototrophic bacteria (i.e., 3.5–4.0 dS/m) [39].

The TOC content in all cases was higher than 20%, the minimum value suggested to increase the content of organic matter in the soil [37]. TC showed the highest TOC content of 32.8%, which was associated with the predominance of organic carbon compounds of difficult degradation in GW, which increased the processing time. There were no significant differences between the TOC of TA and TB ($p = 0.028$); therefore, in this study, bacterial inoculation did not have a significant effect on the TOC of the product. On the other hand, the stability of a product has been identified as a fundamental quality criterion. According to the Rottegrade test, all treatments can be classified as class V and do not represent a potential risk due to the nutrient and oxygen competition between microorganisms in the soil [19].

Regarding the concentration of TN, TB and TC presented the highest concentration values (i.e., 2.2–2.4%) and had significant differences relative to TA (1.32%). However, the TN content for all treatments was higher than 1%, which is the minimum TN value required by Chilean (NCH2880) and Colombian (NTC 5167) regulations. Regarding the maturity of the product, the GI allows for the detection of potential inhibitory effects on seed germination. The TA and TB treatments had GI values greater than 80%. The GI of TA (98.85 ± 2.9) was the highest, probably due to the increased degradation of lignocellulose and phytotoxic substances stimulated by the bacterial inoculum [40]. Likewise, the presence of mineral additives such as PR can increase the availability of nutrients in a product through sorption mechanisms. However, the GI of the TC treatment was below the recommended value of 80% [37]. This indicates that the product may have had phytotoxic substances that affected the germination of radish seeds.

## 4. Discussion

The thermophilic temperatures reached in TA and TB were consistent with those of previous experiments and those reported for the composting of GW [13,41]. Regarding sanitation conditions, all treatments had temperatures of above 50 °C for three or more consecutive days. In this regard, different authors have indicated that this is favorable for the elimination of seeds and pathogens [42]. The bacterial inoculum affected the TA temperature during the cooling phase. The increase in the temperature gradient of TA has also been reported in other studies associated with the incorporation of bacterial inocula. Furthermore, the temperature in this treatment decreased because of the mineralization and humidification of the organic matter. The cooling phase in TA was reduced by 3 days compared with TB and 15 days compared with TC. However, no significant differences were found between TA and TB ($p$ = 0.72) regarding the reduction in the cooling phase. These results are comparable to those reported by Yu et al. [8] on *Bacillus* sp. and *Aspergillus* sp., with no significant differences in the reduction in the processing time of the inoculated treatment.

With respect to pH, inoculation led to slight increases in pH values due to the volatilization of ammonia caused by lignocellulose degradation. Duan et al. [33] explained that bacterial strains can secrete enzymes such as carboxymethyl cellulose, ammonifying enzymes, and xylanase that can affect pH. This result is similar to that reported by Yu et al. [8]. At the end of the process studied here, the average pH values were between 8 and 9, which were in the range reported in another study (between 7 and 9) [42].

Consistent with the temperature results, the highest level of TOC degradation was found in TA and TB throughout the process. However, the greatest amount of degradation was observed in TA after the incorporation of the bacterial inoculum. This indicated that *Bacillus* sp. and *Paenibacillus* sp. could stimulate the enzymatic activity, mineralization, and humidification processes [43]. Similar results were reported by Jiang et al. [18], who used *Trichoderma* sp. and effective microorganisms. Nevertheless, research with lignocellulosic waste has shown that the inoculum does not have a significant effect on the degradation of the TOC and lignocellulose [44]. The differences in the TOC may be due to different distributions of organic components in the different waste sources and differences in local microbial communities [45]. Vrsanska et al. [46] reported that the enzymatic activity during the process depends on the availability of TN, so at the end of the process, its concentration may decrease. In reports using substrates such as cattle manure with the incorporation of bacterial inocula, similar trends have been observed regarding decreases in TN concentration [47]. However, researchers have also reported that inocula do not adversely affect the concentration of TN but rather increase the inorganic nitrogen [44].

On the other hand, the results obtained in this pilot-scale study (120 kg) had some similarities to what was observed in laboratory experiments (20 kg reactors) [10]. In the case of pH, there were no significant differences between both experimental scales for the product obtained in all treatments. A similar result was reported in the composting of GW with a microbial compound inoculum after a scale-up from the laboratory scale (0.05 L reactor) to 1 $m^3$ conical piles [8]. The EC values of the products obtained in both experimental scales of this study were lower than 3 dS/m. However, the pilot-scale values (1.3–1.5 dS/m) were lower than those of the laboratory experiment (1.8–1.9 dS/m) due to the greater leaching of salts in the pilot-scale piles compared with the laboratory reactors that were closed vessels.

At both scales, the lowest TN concentration was found in TA and the highest was found in TC. In the case of TA, this result was probably associated with the fact that the inoculum consumed TN for mineralization processes [40]. In contrast, the greater availability of leaves and lignocellulosic material in TC facilitated the retention of TN at the end of the process. However, this treatment also presented a lower degree of organic matter stability. On the other hand, higher values of TN (1.7–2.4%) were found at the pilot scale compared with the laboratory scale (1.3–1.5%) due to the potential environmental variations of the piles, which could have allowed for the development of different microbial consortia or nitrogen-fixing bacteria that contributed to the conservation of nitrogen in the compost [32]. In contrast, due to their experimental configuration, the 20 kg reactors used in the laboratory had more homogeneous environmental conditions. At both scales, the TOC was lower for the treatments with the substrate mixture (TA and TB) compared with TC. Inoculation did not affect the TOC concentration of the product.

The highest GI values were obtained for the TA product at both scales, with significant differences relative to the uninoculated TB and TC treatments. At the laboratory scale, a GI value of 98.9% was obtained, while at the pilot scale, a value of 95.8% was obtained. However, both scales showed no significant differences in the GI of the TA product.

## 5. Conclusions

The treatment with a substrate mixture and inoculation during the cooling phase (TA) allowed for a reduction of between 4 and 13 days in processing time compared with the treatment with a substrate mixture and no inoculation (TB) and the treatment with only GW (TC). The results indicated that the bacterial inoculum could affect the native microbiota of the process, although a significant reduction in the process was not achieved. TA showed the highest level of lignocellulose biodegradation (31.7%). The final product of TA was the one with the best agricultural characteristics with a pH of 8.3, TOC of 24.8%, TN of 1.32%, and GI of 98.8%. On the other hand, the scaling effect with the incorporation of the bacterial inoculum was shown to affect parameters such as the TOC, TN, GI, and, to a lesser extent, temperature and pH. The differences between scales were due to diffusion and advent transport phenomena. Likewise, the quality of the product found in this pilot-scale study and that previously obtained from the laboratory experiment with the same treatments had similar values of parameters such as the pH, CE, TOC, and GI. However, the greatest difference was found regarding TN, with higher values for the pilot-scale study. The results obtained in this paper highlight the importance of optimizing the composting of GW, specifically with the use of co-substrates and specific inocula, which can be of interest for composting materials with a high content of lignocellulose such as GW.

**Author Contributions:** All authors contributed to the study conception and design. Material preparation, data collection and analysis were performed by E.R.O.-O., J.S.-P., V.S.-T. and A.S. The first draft of the manuscript was written by J.S.-P. and all authors commented on previous versions of the manuscript. All authors have read and agreed to the published version of the manuscript.

**Funding:** This work was funded and supported by Universidad Industrial de Santander under grant No VIE-UIS 2696 and VIE-UIS 2726.

**Conflicts of Interest:** The authors declare no conflict of interest.

## References

1. Wang, W.; Zhang, L.; Sun, X. Improvement of Two-Stage Composting of Green Waste by Addition of Eggshell Waste and Rice Husks. *Bioresour. Technol.* **2021**, *320*, 124388. [CrossRef] [PubMed]
2. Zhang, L.; Sun, X. Addition of Fish Pond Sediment and Rock Phosphate Enhances the Composting of Green Waste. *Bioresour. Technol.* **2017**, *233*, 116–126. [CrossRef]
3. Cui, H.; Ou, Y.; Wang, L.X.; Yan, B.W.; Li, Y.X.; Ding, D.W. Phosphate Rock Reduces the Bioavailability of Heavy Metals by Influencing the Bacterial Communities during Aerobic Composting. *J. Integr. Agric.* **2021**, *20*, 1137–1146. [CrossRef]

4. Reyes-Torres, M.; Oviedo-Ocaña, E.R.; Dominguez, I.; Komilis, D.; Sánchez, A. A Systematic Review on the Composting of Green Waste: Feedstock Quality and Optimization Strategies. *Waste Manag.* **2018**, *77*, 486–499. [CrossRef] [PubMed]
5. Chaher, N.E.H.; Chakchouk, M.; Nassour, A.; Nelles, M.; Hamdi, M. Potential of Windrow Food and Green Waste Composting in Tunisia. *Environ. Sci. Pollut. Res.* **2021**, *28*, 46540–46552. [CrossRef] [PubMed]
6. Torrijos, V.; Calvo Dopico, D.; Soto, M. Integration of Food Waste Composting and Vegetable Gardens in a University Campus. *J. Clean. Prod.* **2021**, *315*, 128175. [CrossRef]
7. Usmani, Z.; Sharma, M.; Karpichev, Y.; Pandey, A.; Chandra Kuhad, R.; Bhat, R.; Punia, R.; Aghbashlo, M.; Tabatabaei, M.; Gupta, V.K. Advancement in Valorization Technologies to Improve Utilization of Bio-Based Waste in Bioeconomy Context. *Renew. Sustain. Energy Rev.* **2020**, *131*, 109965. [CrossRef]
8. Yu, K.; Sun, X.; Li, S.; Cai, L.; Zhang, P.; Kang, Y.; Yu, Z.; Tong, J.; Wang, L. Application of Quadratic Regression Orthogonal Design to Develop a Composite Inoculum for Promoting Lignocellulose Degradation during Green Waste Composting. *Waste Manag.* **2018**, *79*, 443–453. [CrossRef]
9. Rastogi, M.; Nandal, M.; Khosla, B. Microbes as Vital Additives for Solid Waste Composting. *Heliyon* **2020**, *6*, e03343. [CrossRef]
10. Oviedo-Ocaña, E.R.; Soto-Paz, J.; Sanchez-Torres, V.; Castellanos-suarez, L.J.; Komilis, D. Effect of the Addition of the Bacillus Sp., Paenibacillus Sp. Bacterial Strains on the Co-Composting of Green and Food Waste. *J. Environ. Chem. Eng.* **2022**, *10*, 107816. [CrossRef]
11. Grady, E.N.; MacDonald, J.; Liu, L.; Richman, A.; Yuan, Z.C. Current Knowledge and Perspectives of Paenibacillus: A Review. *Microb. Cell. Fact.* **2016**, *15*, 203. [CrossRef] [PubMed]
12. Soto-Paz, J.; Oviedo-Ocaña, E.R.; Angarita-Rangel, M.A.; Rodríguez-Flórez, L.V.; Castellanos-Suarez, L.J.; Nabarlatz, D.; Sanchez-Torres, V. Optimization of Lignocellulolytic Bacterial Inoculum and Substrate Mix for Lignocellulose Degradation and Product Quality on Co-Composting of Green Waste with Food Waste. *Bioresour. Technol.* **2022**, *359*, 127452. [CrossRef]
13. Hernández-Gómez, A.; Calderón, A.; Medina, C.; Sanchez-Torres, V.; Oviedo-Ocaña, E.R. Implementation of Strategies to Optimize the Co-Composting of Green Waste and Food Waste in Developing Countries. A Case Study: Colombia. *Environ. Sci. Pollut. Res.* **2020**, *28*, 24321–24327. [CrossRef] [PubMed]
14. Yu, K.; Li, S.; Sun, X.; Kang, Y. Maintaining the Ratio of Hydrosoluble Carbon and Hydrosoluble Nitrogen within the Optimal Range to Accelerate Green Waste Composting. *Waste Manag.* **2020**, *105*, 405–413. [CrossRef] [PubMed]
15. Asses, N.; Farhat, A.; Cherif, S.; Hamdi, M.; Bouallagui, H. Comparative Study of Sewage Sludge Co-Composting with Olive Mill Wastes or Green Residues: Process Monitoring and Agriculture Value of the Resulting Composts. *Process Saf. Environ. Prot.* **2018**, *114*, 25–35. [CrossRef]
16. Rawoteea, S.A.; Mudhoo, A.; Kumar, S. Co-Composting of Vegetable Wastes and Carton: Effect of Carton Composition and Parameter Variations. *Bioresour. Technol.* **2017**, *227*, 171–178. [CrossRef]
17. Hemidat, S.; Jaar, M.; Nassour, A.; Nelles, M. Monitoring of Composting Process Parameters: A Case Study in Jordan. *Waste Biomass Valorization* **2018**, *9*, 2257–2274. [CrossRef]
18. Jiang, J.; Liu, X.; Huang, Y.; Huang, H. Inoculation with Nitrogen Turnover Bacterial Agent Appropriately Increasing Nitrogen and Promoting Maturity in Pig Manure Composting. *Waste Manag.* **2015**, *39*, 78–85. [CrossRef]
19. Lü, F.; Shao, L.M.; Zhang, H.; Fu, W.D.; Feng, S.J.; Zhan, L.T.; Chen, Y.M.; He, P.J. Application of Advanced Techniques for the Assessment of Bio-Stability of Biowaste-Derived Residues: A Minireview. *Bioresour. Technol.* **2018**, *248*, 122–133. [CrossRef]
20. Komilis, D.P.; Tziouvaras, I.S. A Statistical Analysis to Assess the Maturity and Stability of Six Composts. *Waste Manag.* **2009**, *29*, 1504–1513. [CrossRef]
21. Sluiter, A.; Hames, B.; Ruiz, R.; Scarlata, C.; Sluiter, J.; Templeton, D.; Nrel, D.C. *Determination of Structural Carbohydrates and Lignin in Biomass, in Laboratory Analyticial Procedure (LAP)*; National Renewable Energy Laboratory: Golden, CO, USA, 2012.
22. Oviedo-ocaña, E.R.; Hernández-gómez, A.M.; Ríos, M.; Portela, A.; Sánchez-torres, V.; Domínguez, I.; Komilis, D. A Comparison of Two-stage and Traditional Co-composting of Green Waste and Food Waste Amended with Phosphate Rock and Sawdust. *Sustainability* **2021**, *13*, 1109. [CrossRef]
23. Wang, H.; Zhao, Y.; Wei, Y.; Zhao, Y.; Lu, Q.; Liu, L.; Jiang, N.; Wei, Z. Biostimulation of Nutrient Additions on Indigenous Microbial Community at the Stage of Nitrogen Limitations during Composting. *Waste Manag.* **2018**, *74*, 194–202. [CrossRef] [PubMed]
24. Zeng, G.; Yu, M.; Chen, Y.; Huang, D.; Zhang, J.; Huang, H.; Jiang, R.; Yu, Z. Effects of Inoculation with Phanerochaete Chrysosporium at Various Time Points on Enzyme Activities during Agricultural Waste Composting. *Bioresour. Technol.* **2010**, *101*, 222–227. [CrossRef] [PubMed]
25. Cáceres, R.; Coromina, N.; Malińska, K.; Martínez-Farré, F.X.; López, M.; Soliva, M.; Marfà, O. Nitrification during Extended Co-Composting of Extreme Mixtures of Green Waste and Solid Fraction of Cattle Slurry to Obtain Growing Media. *Waste Manag.* **2016**, *58*, 118–125. [CrossRef]
26. Zhou, C.; Liu, Z.; Huang, Z.-L.; Dong, M.; Yu, X.-L.; Ning, P. A New Strategy for Co-Composting Dairy Manure with Rice Straw: Addition of Different Inocula at Three Stages of Composting. *Waste Manag.* **2015**, *40*, 38–43. [CrossRef]
27. Xi, B.; He, X.; Dang, Q.; Yang, T.; Li, M.; Wang, X.; Li, D.; Tang, J. Effect of Multi-Stage Inoculation on the Bacterial and Fungal Community Structure during Organic Municipal Solid Wastes Composting. *Bioresour. Technol.* **2015**, *196*, 399–405. [CrossRef] [PubMed]

28. Wu, D.; Wei, Z.; Qu, F.; Mohamed, T.A.; Zhu, L.; Zhao, Y.; Jia, L.; Zhao, R.; Liu, L.; Li, P. Effect of Fenton Pretreatment Combined with Bacteria Inoculation on Humic Substances Formation during Lignocellulosic Biomass Composting Derived from Rice Straw. *Bioresour. Technol.* **2020**, *303*, 122849. [CrossRef]
29. Bohacz, J. Lignocellulose-Degrading Enzymes, Free-Radical Transformations during Composting of Lignocellulosic Waste and Biothermal Phases in Small-Scale Reactors. *Sci. Total Environ.* **2017**, *580*, 744–754. [CrossRef] [PubMed]
30. Zhou, S.-P.; Zhou, H.-Y.; Xia, S.-N.; Ying, J.-M.; Ke, X.; Zou, S.-P.; Xue, Y.-P.; Zheng, Y.-G. Efficient Bio-Degradation of Food Waste through Improving the Microbial Community Compositions by Newly Isolated Bacillus Strains. *Bioresour. Technol.* **2021**, *321*, 124451. [CrossRef]
31. Feng, C.; Zeng, G.; Huang, D.; Hu, S.; Zhao, M.; Lai, C.; Huang, C.; Wei, Z.; Li, N. Effect of Ligninolytic Enzymes on Lignin Degradation and Carbon Utilization during Lignocellulosic Waste Composting. *Process Biochem.* **2011**, *46*, 1515–1520. [CrossRef]
32. Song, C.; Li, M.; Wei, Z.; Jia, X.; Xi, B.; Liu, D.; Zhu, C.; Pan, H. Effect of Inoculation with Multiple Composite Microorganisms on Characteristics of Humic Fractions and Bacterial Community Structure during Biogas Residue and Livestock Manure Co-Composting. *J. Chem. Technol. Biotechnol.* **2016**, *91*, 155–164. [CrossRef]
33. Duan, M.; Zhang, Y.; Zhou, B.; Qin, Z.; Wu, J.; Wang, Q.; Yin, Y. Effects of *Bacillus Subtilis* on Carbon Components and Microbial Functional Metabolism during Cow Manure–Straw Composting. *Bioresour. Technol.* **2020**, *303*, 122868. [CrossRef]
34. Kausar, H.; Ismail, M.R.; Saud, H.M.; Othman, R.; Habib, S. Use of Lignocellulolytic Microbial Consortium and Ph Amendment on Composting Efficacy of Rice Straw. *Compost. Sci. Util.* **2013**, *21*, 121–133. [CrossRef]
35. Oviedo-Ocaña, E.R.; Dominguez, I.; Komilis, D.; Sánchez, A. Co-Composting of Green Waste Mixed with Unprocessed and p Food Waste: Influence on the Composting Process and Product Quality. *Waste Biomass Valorization* **2019**, *10*, 63–74. [CrossRef]
36. Khatibi, H.; Hassani, A. Effective Management and Composting of Organic Wastes Using New Developed Consortia. *Environ. Dev. Sustain.* **2021**, *23*, 16891–16910. [CrossRef]
37. Lasaridi, K.; Protopapa, I.; Kotsou, M.; Pilidis, G.; Manios, T.; Kyriacou, A. Quality Assessment of Composts in the Greek Market: The Need for Standards and Quality Assurance. *J. Environ. Manag.* **2006**, *80*, 58–65. [CrossRef]
38. Bernal, M.P.; Sommer, S.G.; Chadwick, D.; Qing, C.; Guoxue, L.; Michel, F.C. Chapter Three—Current Approaches and Future Trends in Compost Quality Criteria for Agronomic, Environmental, and Human Health Benefits. In *Advances in Agronomy*; Sparks, D.L., Ed.; Academic Press: Cambridge, MA, USA, 2017; Volume 144, pp. 143–233.
39. Manu, M.K.; Kumar, R.; Garg, A. Performance Assessment of Improved Composting System for Food Waste with Varying Aeration and Use of Microbial Inoculum. *Bioresour. Technol.* **2017**, *234*, 167–177. [CrossRef]
40. Chen, R.; Senbayram, M.; Blagodatsky, S.; Myachina, O.; Dittert, K.; Lin, X.; Blagodatskaya, E.; Kuzyakov, Y. Soil C and N Availability Determine the Priming Effect: Microbial N Mining and Stoichiometric Decomposition Theories. *Glob. Change Biol.* **2014**, *20*, 2356–2367. [CrossRef]
41. Mishra, S.K.; Yadav, K.D. Application of Locally Available Microbial Inoculant to Accelerate Green Waste Composting at a Community Level. *Bioresour. Technol. Rep.* **2021**, *16*, 100859. [CrossRef]
42. Wan, L.; Wang, X.; Cong, C.; Li, J.; Xu, Y.; Li, X.; Hou, F.; Wu, Y.; Wang, L. Effect of Inoculating Microorganisms in Chicken Manure Composting with Maize Straw. *Bioresour. Technol.* **2020**, *301*, 122730. [CrossRef]
43. Vu, V.; Farkas, C.; Riyad, O.; Bujna, E.; Kilin, A.; Sipiczki, G.; Sharma, M.; Usmani, Z.; Gupta, V.K.; Nguyen, Q.D. Enhancement of the Enzymatic Hydrolysis Efficiency of Wheat Bran Using the Bacillus Strains and Their Consortium. *Bioresour. Technol.* **2022**, *343*, 126092. [CrossRef] [PubMed]
44. Nigussie, A.; Dume, B.; Ahmed, M.; Mamuye, M.; Ambaw, G.; Berhiun, G.; Biresaw, A.; Aticho, A. Effect of Microbial Inoculation on Nutrient Turnover and Lignocellulose Degradation during Composting: A Meta-Analysis. *Waste Manag.* **2021**, *125*, 220–234. [CrossRef] [PubMed]
45. Ma, C.; Lo, P.K.; Xu, J.; Li, M.; Jiang, Z.; Li, G.; Zhu, Q.; Li, X.; Leong, S.Y.; Li, Q. Molecular Mechanisms Underlying Lignocellulose Degradation and Antibiotic Resistance Genes Removal Revealed via Metagenomics Analysis during Different Agricultural Wastes Composting. *Bioresour. Technol.* **2020**, *314*, 123731. [CrossRef]
46. Vrsanska, M.; Buresova, A.; Damborsky, P.; Adam, V. Influence of Different Inducers on Ligninolytic Enzyme Activities. *J. Met. Nanotechnologies* **2015**, *3*, 64–70.
47. Li, J.; Wang, X.; Cong, C.; Wan, L.; Xu, Y.; Li, X.; Hou, F.; Wu, Y.; Wang, L. Inoculation of Cattle Manure with Microbial Agents Increases Efficiency and Promotes Maturity in Composting. *3 Biotech.* **2020**, *10*, 128. [CrossRef] [PubMed]

processes

MDPI

Article

# Effect of Poultry Manure-Derived Compost on the Growth of *eucalypts* spp. Hybrid Clones

**Pedro F. Rizzo [1,*], María C. Salinas [1,2], Virginia Della Torre [3], Juan P. Diez [4], Leonardo F. Sallesses [5], Nicolás I. Riera [3], Pablo S. Pathauer [6], Dimitrios Komilis [7] and Antoni Sánchez [7,*]**

1 Instituto Nacional de Tecnología Agropecuaria (INTA), Estación Experimental Agropecuaria Mendoza, Agrotecnología Sostenible, Luján de Cuyo 5507, Argentina
2 National Council of Scientific and Technical Research (CONICET), Mendoza 5500, Argentina
3 Laboratorio de Transformación de Residuos, Instituto Nacional de Tecnología Agropecuaria (INTA), Instituto de Microbiología y Zoología Agrícola (IMyZA), Hurlingham 1686, Argentina
4 Instituto Nacional de Tecnología Agropecuaria, Estación Experimental Agropecuaria Bariloche, Rio Negro, San Carlos de Bariloche 8400, Argentina
5 Instituto Nacional de Tecnología Agropecuaria (INTA), Estación Experimental Agropecuaria Balcarce, Balcarce 7620, Argentina
6 Instituto Nacional de Tecnología Agropecuaria (INTA), Instituto de Recursos Biológicos (IRB), Buenos Aires 1033, Argentina
7 Composting Research Group, Department of Chemical Engineering, Universitat Autònoma de Barcelona, Bellaterra, 08193 Cerdanyola del Vallès, Spain
* Correspondence: rizzo.pedro@inta.gob.ar (P.F.R.); antoni.sanchez@uab.cat (A.S.); Tel.: +54-(261)-4963020 (P.F.R.); +34-935811019 (A.S.)

**Citation:** Rizzo, P.F.; Salinas, M.C.; Della Torre, V.; Diez, J.P.; Sallesses, L.F.; Riera, N.I.; Pathauer, P.S.; Komilis, D.; Sánchez, A. Effect of Poultry Manure-Derived Compost on the Growth of *eucalypts* spp. Hybrid Clones. *Processes* **2022**, *10*, 2182. https://doi.org/10.3390/pr10112182

Academic Editor: Davide Dionisi

Received: 4 October 2022
Accepted: 21 October 2022
Published: 25 October 2022



**Abstract:** Interspecific hybrids of *E. grandis* × *E. camaldulensis* were generated to widen the plantation area. The aim of this study was to assess root capability and development for six different clones of eucalyptus grown in substrates made with three different composts derived from poultry manure. A factorial design was used to assess the effect of different composts on six growth variables. The analysis detected a greater effect from the genotype than the substrate. *E. grandis* × *E. camaldulensis* hybrid vegetative propagation was successful in alternative substrates formulated from composted poultry manure. GC8 was the genotype that showed the greatest differences for four the different variables among the substrates, being both the most sensitive and the one with the highest values for all parameters measured. The hybrids' vegetative propagation was determined in alternative substrates formulated from poultry manure compost. The physicochemical characteristics of substrates composed of pine bark and sawdust provided adequate conditions for the growth of eucalyptus. GC8 was the genotype most sensitive to the use of different substrates, showing significant differences in the ratio of roots/callus, radicular dry weight, and cutting dry weight. These clones might be a good option for evaluating compost-based substrates for forestry applications.

**Keywords:** substrates; vegetative propagation; forest valorization

## 1. Introduction

Eucalyptus is one of the most important economic forest species worldwide [1,2]. *E. grandis* has fast growth and light wood; however, it is sensitive to frost, and it is mainly destined for solid use. *E. camaldulensis* is a slow growth species with a dark and dense wood. It has wide genetic variation and plasticity, which allows it to adapt to different climatic and soil conditions, including those with high salinity and low moisture [3]. Hybrids are able to produce adaptive features and production combinations which are not genetically possible within species.

A hybrid can only be used if it can be cloned, and hybrid vigor is achieved using clonal silviculture [4]. In this sense, *E. grandis* and *E. camaldulensis*, have the advantage of easy rooting if variables that intervene in their vegetative propagation are adjusted, such as humidity, temperature, rooting promoter concentration, and substrate, among others. The substrate is the material that allows the radicular system anchorage [5]. It can be made up of one or more materials, and must have high porosity and hydric retention capacity as well as good drainage and aeration [6]. Different materials such as composted pine bark, peat, perlite, and burnt rice bran are often used, either pure or mixed and in different proportion, as substrates in eucalyptus vegetative propagation.

Developing peat alternative substrates is necessary for three different reasons: (i) the sources of peat are limited worldwide; (ii) the pressure for using waste coming from human or industrial activities is increasing rapidly; and (iii) the economic necessity of using locally produced waste products is increasing [7]. The most commonly used material in substrate mixture for forest plant production in nurseries is peat moss, thanks to its physicochemical characteristics [8]. This material comes from mosses, such as *Sphagnum*; however, its use is being debated because of its high cost [9] and its questionable future availability due to environmental limitations [10]. The substrate mixture should favor well-developed fibrous root systems to produce quality plants and improve their survival and growth in the field [11]. To achieve the optimal conditions of the substrate, the mixtures should have adequate physical characteristics to retain water and facilitate drainage and aeration [12,13].

The use of organic waste derived from agricultural and forestry activities is becoming important for plant production [4,14]. Pine sawdust and bark are wood waste products of the forestry industry, and can be obtained at a relatively low cost [15]. However, there is a lack of information on the possibility of commercial use of these growing mediums in forest nurseries [16]. In substrate mixtures, different percentages are used depending on the compost properties and the species under cultivation. Generally, these mixtures have alkaline pH and high salt content. For this reason, it is convenient to combine them with materials with lower pH and salt levels [17].

A consequence of the amplification of animal production systems worldwide is the concentration of animals in small areas, resulting in large amounts of manures and excreta. This leads to environmental problems, including water and soil pollutions and bad odors around animal breeding sites [18–20]. Common treatment alternatives for poultry manure are anaerobic digestion and composting or a combination of both [21–23].

Argentinian eucalyptus and poultry production are mainly concentrated in the same region of the Littoral, namely, Entre Rios, one of the most important provinces [24]. This makes it an ideal area to study strategies for using poultry manure as a potential substrate component for substrates in eucalyptus plantations. The use of a problematic waste as an alternative to peat for substrate formulation has multiple benefits, for example, recovery of poultry manure, improving the biogeochemical cycling of nutrients, and minimizing contamination [25]. For this reason, the aim of this study was to evaluate the effect of different substrates which included poultry manure derived compost in the development of six different eucalyptus hybrid clones.

## 2. Materials and Methods

### 2.1. Production of Compost and Substrates

The compost was made in the Laboratorio de Transformación de Residuos (LTR) IMyZA, CICVyA—INTA using manure from hens in automatized production facilities. The three composts (C1, C2, C3) used to formulate the substrates (S1, S2, S3) had different initial compositions. Table 1 shows the percentage composition of the three composts [25]. Sawdust and bark have different physical characteristics; sawdust has a larger percentage of fine particles and can cause problems with excess moisture, while wood shavings have larger particles and a low water holding capacity [26]. Crushed bare corn cobs combine fine particles and large ones.

**Table 1.** Initial waste percentages (%) [1] used in each compost to formulate the substrates (S1, S2, and S3).

| Type of Compost | Poultry Manure | Intact Corn Bare Cobs | Crushed Corn Bare Cobs | Sawdust | Wood Savings |
|---|---|---|---|---|---|
| C1 [2] | 40 | 0 | 20 | 20 | 20 |
| C2 [3] | 60 | 0 | 20 | 20 | 0 |
| C3 [4] | 60 | 30 | 10 | 0 | 0 |

[1] All percentages are on a wet weight basis. [2] Compost 1. [3] Compost 2. [4] Compost 3.

Poultry manure is rich in nitrogen waste and can be mixed with materials rich in available carbon, such as sawdust. Sawdust can retain nitrogen, thereby avoiding ammonia volatilization, because organisms that decompose organic matter use nitrogen as well. Poultry manure contains a low C:N ratio and high porosity. Therefore, degradation processes can be enhanced by adding carbon-rich materials (co-substrates). The co-composting of poultry manure with other agricultural wastes improves its physicochemical characteristics and reduces phytotoxicity, in addition to promoting and managing better use of other local residues, thereby generating added value. In this sense, we were able to produce compost with 40–60% poultry manure and other agricultural wastes as co-substrates [25].

Three mixtures were prepared using a capacity of 0.5 $m^3$ mix at the beginning of the assay. Piles were built in trapezoidal shapes (1.5 m high, 2 m wide and 2 m length). Each treatment was carried out using three repetitions of 2 $m^3$ piles each. The composting piles were manually turned every three days during the first active decomposition phase of the process and every five days when the cell temperature was similar to the ambient temperature. Moisture content was maintained through irrigation and taking into account local precipitation. The composting process lasted 83 days. All the composts reached the stage of stability and maturity as measured from the static respirometric index (SRI < 0.5 mg $O_2$ $g^{-1}$ OM $h^{-1}$) and the $NH_4^+/NO_3^-$ ratio (<0.3), respectively (Table 2). Table 2 shows the main parameters of the three composts.

The substrates used in this study (S1, S2, and S3) were made of 40% of compost 1 (C1), compost 2 (C2), or compost 3 (C3) and 60% composted pine leaves (Table 3). These formulations were defined taking into account previous pH and electrical conductivity measurements of the substrates that had been prepared with different proportions of composts and composted pine leaves. In addition, we took into account the fact that the optimum pH for substrates used in plant production should be between 5.2 and 6.3, while a value below 4.0 can cause root disease [6]. While livestock manure compost has good physical properties, the soluble salt content is too high and the pH is alkaline [27]. For this reason, it is necessary to mix it with an acidic material, such as peat, or with compost with high cationic exchange capacity [7]. The control substrate (C) was made with a 50:50 (*v/v*) mixture of composted pine bark and burned rice husks, which is routinely used for vegetative propagation and is recommended by various authors [28,29].

**Table 2.** Physicochemical characterization of composts used in the substrate preparation.

| Parameters [1] | Unit | C1 | C2 | C3 | Target Value or Range/Upper Limit | Reference |
|---|---|---|---|---|---|---|
| pH | | 8.0 | 8.1 | 8.7 | 6–8/9 | [30] |
| EC | mS $cm^{-1}$ | 2.0 | 3.4 | 2.7 | <0.6/1.5 | [30] |
| D | Mg $m^{-3}$ | 0.7 | 0.8 | 0.5 | 0.45–0.50/0.55 | [30] |
| OM | % | 35.0 | 37.0 | 87.0 | ≥20 | [31] |
| SRI | mg $O_2$ $g^{-1}$ OM $h^{-1}$ | 0.20 | 0.40 | 0.37 | 0.5–1.0 | [32] |
| C:N | % | 14.4 | 13.6 | 18.0 | ≤20 | [31] |
| Ca | mg $L^{-1}$ | 8170 | 9810 | 2210 | ≥1% | [33] |
| Mg | mg $L^{-1}$ | 251 | 348 | 470 | ≥0.05% | [33] |
| K | mg $L^{-1}$ | 6563 | 8700 | 11,350 | NPK ≥ 6% | [33] |
| $NH_4^+/NO_3^-$ | % | 0.0 | 0.0 | 0.004 | <0.3 | [31] |

[1] EC: electrical conductivity, D: apparent density, OM: organic matter, SRI: static respirometric index, C:N: carbon nitrogen ratio, Ca: calcium, Mg: magnesium, K: potassium, $NH_4^+/NO_3^-$: ammonium nitrate ratio.

**Table 3.** Initial substrate formulation, pH, and EC values (mean ± standard deviation with $n$ = 3).

| Substrates | Composition [1] | pH | EC * |
|---|---|---|---|
| 1 | 40% Compost 1 + 60% composted pine leaves | 6.55 ± 0.09 | 871 ± 7.1 |
| 2 | 40% Compost 2 + 60% composted pine leaves | 6.64 ± 0.17 | 1387 ± 121.1 |
| 3 | 40% Compost 3 + 60% composted pine leaves | 7.03 ± 0.06 | 1956 ± 48.8 |
| Control | 50% burned rice husks + 50% composted pine bark | 7.05 ± 0.02 | 653 ± 1.1 |

[1] The percentages are on a volume basis. The control substrate is routinely used for vegetative propagation in the IRB; * Electrical conductivity (μS $cm^{-1}$).

### 2.2. Forest Assessment of Compost and Substrates

The study was carried out at the Instituto de Recursos Biológicos (IRB), CIRN–INTA in Buenos Aires, Argentina. Lopez (2017) [34] indicates that in Argentina the first experimental records of interspecific hybrids were reported by Alliani (1990) [35]. In the Mesopotamia region of Argentina, the first controlled crosses for the selection of hybrid clones of *E. grandis* × *E. camaldulensis* and *E. grandis* × *E. tereticornis* were generated by the INTA clonal program (Harrand and Schenone, 2002) [36]. More recently, Harrand et al. (2016) [37] report that the INTA clonal program has 150 hybrid clones in different stages of evaluation, several of which have been registered in the National Registry of Cultivars (INASE), and have since 2014 been transferred to nurseries in the region through transfer agreements. In this study, four clones of *E. grandis* × *E. camaldulensis* (GC6, GC8, GC19 and GC24) and two clones of *E. grandis* × *E. tereticornis* (5–105 and 5–128) were used, all originating from the program clone of INTA. The cuttings were obtained from 30 to 50 cm long young regrowth of the six genotypes from a cloning garden at the IRB. The cuttings were 10 cm long with an internode and preferably two opposite leaves, with its foliar area reduced by 50% to minimize evapotranspiration. Each cutting apical part was cut in a straight shape, while the basal extreme was cut in beveled edge to originate a greater contact zone with the growth promoter. Then, the cuttings were put under a Captan® (2 g $L^{-1}$) fungicide solution for one minute and treated at their bases with indolbutiric acid 3000 ppm dispersed in industrial talcum powder. Then, they were implanted in individual 145 $cm^3$ substrate capacity tubes.

The trial was located within a 100 μ crystal polyethylene greenhouse with an automatized irrigation and relative humidity control system. Environmental humidity values were near 80%, and environmental temperature ranged between 22 °C and 30 °C. The trial cuttings were sprinkled weekly with COMBO brand (2.5 $cm^3$ $L^{-1}$) complete fertilizing solution with foliar absorption mixed with Captan® (2 g $L^{-1}$). The fungicide was altered weekly with Carbendazim® (2 $cm^3$ $L^{-1}$) to avoid fungal resistance.

From the 55 days from the trial beginning, the humidification system was stopped to simulate environmental conditions until the plants were removed from the substrates and their roots washed. Then, the live cuttings were divided in three fractions (shoots, cuttings, and roots) and dried in a 50 °C oven until constant weight was reached. Each fraction was weighed separately in a precision balance.

The tested variables to assess the growth of the eucalypts clones were survival (SUR), defined as the proportion of live cuttings from five of each plot, primary leave number (PLN), average number of shoots (ANS), average amount of leaves per shoot (ALS), total leaves (TL), average shoot length (ASL), longest shoot length (LSL),roots/callus (R/C), root length (RL), cutting diameter (CD), radicular dry weight (RDW), shoot dry weight (SDW), cutting dry weight (CDW), root type (RT), and shoot type (ST).

To evaluate the physicochemical substrate and compost characteristics, the following parameters were measured in the substrates according to USDA and USCC (2001) [32] and INTA (2021) [38] (Tables 2 and 3): pH, electrical conductivity (EC), humidity, dry matter, organic matter, ashes, total organic carbon (TOC), and total nitrogen (TN).

### *2.3. Experimental Design and Statistical Analysis*

A factorial experimental design was used in which four substrates and six clones were tested. Eight repetitions for each combination of substrate and clone were analyzed, and five eucalyptus cuttings were used in each repetition.

A principal component analysis (PCA) was carried out to study the multivariate differences between the clones grown in different substrates. Subsequently, a Kruskal–Wallis non-parametric test was chosen to study the differences between the substrates for each clone and for each variable in particular ($p \leq 0.05$). All the statistical analyses were run using InfoStat [39].

## 3. Results

### *3.1. Principal Component Analysis*

PCA component 1 explained 57.1% of the variability, while component 2 explained 11.1%. Component 1 allowed us to group different eucalypts genotypes. The highest vegetable growth values were associated with GC8 and GC19, whereas the others clones were associated with lower vegetable growth values (Figure 1). The average number of shoots (ANS) is the variable that best explained Component 2, although there were no significant differences within the treatments in a later analysis within this particular variable.

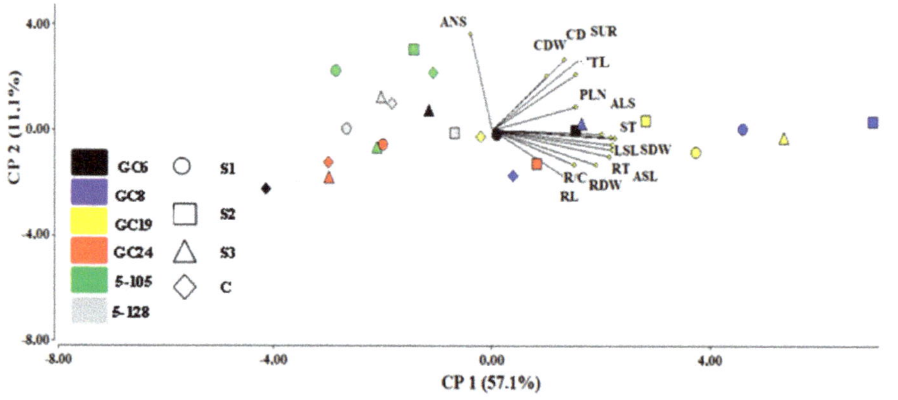


**Figure 1.** Principal component analysis plot. References: ANS (Average number of shoots), CDW (cutting dry weight), CD (cutting diameter), SUR (survival), TL (total leaves), PLN (primary leave number), ALS (average amount of leaves per shoot), ST (shoot type), LSL (longest shoot length), SDW (shoot dry weight), RT (root type), ASL (average shoot length), R/C (roots/callus), RDW (radicular dry weight), RL (root length).

The analysis showed that most of the variables were actually correlated. The ones considered for the analysis were those that had correlations with ten or less variables, namely, survival, number of shoots per cutting, total leaves per cutting, roots number/callus ratio, root length, cutting diameters, dried root weight, and dried cutting weight.

### *3.2. Substrates Properties in Eucalypts Nurseries*

Table 3 shows that the initial pH values are close to the optimum range cited in the bibliography [40]. Substrate 3 and control showed values significantly higher than the remaining substrate. On the other hand, the EC values were significantly higher in those substrates with larger amounts of poultry manure in their initial composition (substrates 2 and 3).

### 3.3. *Growth Variables of the Clones Evaluated*

#### 3.3.1. Survival Rates

Table 3 shows that the total average survival was 50%. S1 and S3 manifested a 56% survival rate and C showed 41% survival, while S2 showed 47%. GC19 and GC8 showed a survival rate greater than 60%. GC24 had the highest percentage of mortality. The highest survival rate (90%) was found in the combinations of GC8 with S1 and S3 and GC19 with S2 (Table 4).

**Table 4.** Survival Percentages (%).

| Treatments | S1 | S2 | S3 | C | Clone |
|---|---|---|---|---|---|
| GC6 | 60 | 55 | 68 | 8 | 48 |
| GC8 | 88 | 53 | 88 | 25 | 63 |
| GC19 | 60 | 85 | 65 | 50 | 65 |
| GC24 | 45 | 25 | 15 | 35 | 30 |
| 5–105 | 55 | 30 | 60 | 75 | 55 |
| 5–128 | 30 | 35 | 43 | 55 | 41 |
| Substrates | 56 | 47 | 56 | 41 | 50 |

#### 3.3.2. Substrate Effect Analysis Considering Each Genotype

When analyzing the substrate effect, neither the shoots number per cutting or cutting root length showed significant differences in any of the genotypes.

Only GC19 showed significant differences within total leaves. It was higher in S1 and S2 (16 ± 6) than in S3 and C (11 ± 5) (Figure 2). An opposite result was obtained when considering cutting dry weight. GC8 showed significant differences, having a higher CDW with S3 and C (0.3 ± 0.1 g) than with S1 (0.2 ± 0.1 g). Again, only GC8 showed significant differences, in this case considering the roots/callus variable. Figure 2 shows that this ratio was significantly higher in S3 than in the rest of the substrates (S3 = 10 ± 6; S1, S2 and C = 7 ± 6). The effect of the substrates was significant in more clones when analyzing the radicular dry weight. GC8 significantly differed between S2 (with the lowest value 0.02 ± 0.02 g) and S1 and S3 (the highest value 0.04 ± 0.02 g). GC24 significantly differed between S1 and S2 (with the lowest value, 0.01 ± 0.01 g) and S3 (the highest value 0.03 ± 0.02 g). In 5–105 clones, C and S1 had significant differences, with C having the highest RDW (0.02 ± 0.01 gr) and S1 the lowest (0.01 ± 0.01 g). In the case of 5–128, S1 (0.01 ± 0.00 g) differed significantly from the rest of the substrates, which had a greater RDW value (0.02 ± 0.01 g). Finally, when we evaluated the effect of the substrates on the cutting diameter of the clones, we obtained GC6 grown in S1 and S3 showed a significantly greater diameter than the control (S1 and S3 1.7 ± 0.5 mm and C = 1.4 ± 0.4 mm). The diameter in GC8 was significantly higher in S3 (2.5 ± 0.8 mm), medium in S2 (2.0 ± 0.8 mm), and lower in C (1.5 ± 0.7 mm). Those in S1 had an intermediate value between S2 and S3, with no significant differences between them. GC24 clones had significantly larger cutting diameter in C than in S3 (1.6 ± 0.3 mm and 1.4 ± 0.5 mm).

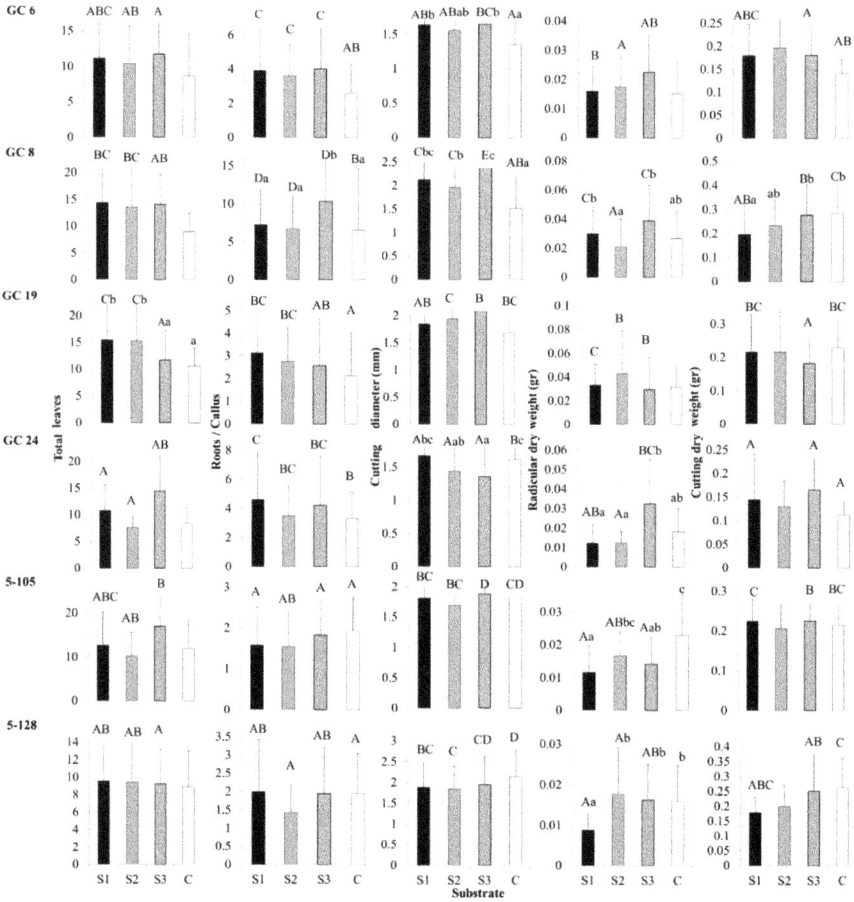

**Figure 2.** Growth variables (Total Leaves, Roots/callus, Cutting Diameter, Radicular Dry Weight, Cutting Dry Weight) in four substrates (1, 2, 3, and control) and six clones (GC6, GC8, GC19, GC24, 5–105 and 5–128). The bars indicate mean values of Total Leaves, Roots/callus, Cutting Diameter, Radicular Dry Weight, and Cutting Dry Weight of eight replicates ± standard deviation. Different lowercase letters indicate significant differences between substrates for each variable analyzed ($p \leq 0.05$). Different capital letters indicate significant differences between genotypes for each variable analyzed ($p \leq 0.05$). For no significant difference, no letters are shown.

#### 3.3.3. Genotype Effect Analysis Considering Each Substrate

When we focused on genotype effect, the shoot number per cutting did not show significant differences in any of the substrates.

S1 presented significant differences in total leaves. GC19 obtained the highest total leaves per cutting (15 ± 7), while GC24 presented the lowest (9 ± 6). Only these two genotypes were different; the rest had intermediate values (Figure 2). The same trend was found using S2 (15 ± 7 for GC19 and 8 ± 2 for GC24). Within S3, the total number of leaves per cutting was significantly higher in 5–105 (17 ± 8) than in 5–128, GC19 and 5–128 (11 ± 5), which had the lowest. There were no differences within the genotypes grown in the control substrate.

Using S1, GC8 showed the highest root number/callus (7 ± 4) and 5–105 the lowest (2 ± 1). The same results were found using S3 (GC8: 10 ± 6 and 5–105: 2 ± 1). In S2, GC8 had the highest ratio value by a significant margin (7 ± 4), while 5–128 had the lowest

(1 ± 1). The control presented significant differences as well. GC8 and GC24 had the highest ratio (4 ± 5) and GC19, 5–105 and 5–128 had the lowest (2 ± 1), while GC6 had a value between these groups.

Regarding root length, there were differences for genotypes grown in S1. GC19 had significantly longer roots than the rest of the genotypes (18.4 ± 6.3 cm), 5–105 had the shortest roots (11.7 ± 2.9 cm), and GC8 had an intermediate value (14.3 ± 4.8 cm). GC6, GC24, and 5–128 had intermediate length between 5–105 and GC8, with no significant differences. GC19 grown in S2 had significantly longer roots (17.8 ± 5.1 cm) than the rest of the genotypes (12.9 ± 4.5 cm). In S3, GC19 had significantly longest root than 5–105 (17.2 ± 6.3 cm and 11.4 ± 3.0 cm, respectively).

Another variable that showed significant differences was the cutting diameter. In S1, the GC8 diameter was significantly higher than that of GC24, which had the lowest value (2.1 ± 0.7 mm and 1.7 ± 0.7 mm, respectively). Using S2, GC8, GC19 and 5–128 had significantly higher diameters (1.9 ± 0.7 mm) than that of GC24 (1.4 ± 0.4 mm). Finally, in S3, GC8 had the highest diameter (2.5 ± 0.8 mm), the diameters of GC6, GC19, 5–105 and 5–128 were grouped in an intermediate level, and the diameter of GC24 was significantly the lowest one (1.4 ± 0.5 mm). GC8 grown in S3 had the highest diameter (2.5 ± 0.8 mm), while GC 24 had the lowest (1.4 ± 0.5 mm). In the control (C), only 5–128 and GC6 differed significantly, with 5–128 having the highest diameter and GC6 having the lowest (2.2 ± 0.6 mm and 1.4 ± 0.4 mm respectively).

Another variable considered was the radicular dry weight. Focusing on S1, GC8 and GC19 had significantly higher RDW than 5–105 and 5–128 (0.03 ± 0.02 g and 0.01 ± 0.01 g respectively). GC6 and GC24 had no significant differences in values between them. In S2, GC19 had a significantly higher RDW than the rest of the substrates (0.04 ± 0.04 g) except 5–105, which showed no significant differences with any of the genotypes. The genotypes with the lowest values were GC6, GC8, GC24, and 5–128 (0.02 ± 0.01 g). In S3, there were significant differences between GC8 and 5–105, the first with the highest RDW and the second with the lowest (0.04 ± 0.02 g and 0.01 ± 0.01 g, respectively); the remaining genotypes had values between these levels. Again, the genotypes showed no differences when using the control substrate.

The last variable considered that showed significant differences was cutting dry weight. For S1, 5–105 had the greatest value (0.2 ± 0.1 g) and GC24 had the lowest ne (0.1 ± 0.1 g). In S3, 5–105 and GC8 (0.3 ± 0.1 g) had significantly higher values than GC6, GC19, and GC24 (0.2 ± 0.1 g). In C, GC8 and 5–128 (0.3 ± 0.1 g) showed heavier CDW than GC24 (0.1 ± 0.0 g).

## 4. Discussion

The differences between the treatments were affected principally by the genotype (clone). Component 1 from the PCA explained 57.1% of the variability. This fact agrees with the findings of Woodward et al. (2006) [41], who highlighted the genetic differences that exist between clones within a species in their investigation. Wherever more than one clone was used in an experiment, there was a significant difference in the number of shoots and roots produced between the clones. The authors highlighted the use of PCA, which allowed them to verify that B availability was the attribute that explained most of eucalypt biomass variation. Additionally, these analyses showed the characteristics that are crucial to improving the organic waste compost-based substrates in comparison to commercial substrates [42]. Rinaldi et al. (2014) [43] used a multivariate approach to data analysis and summarized the information of a complex substrate–plant system based on a large set of tested treatments and several observed variables. Their results suggest that a combination of indicators is needed to evaluate the physiological response of ornamental rosemary to the different treatments explored and to classify treatments into homogeneous groups based on plant response. They concluded that multivariate analysis evaluates the influence of different composts on the substrate properties and the complexity of plant response better than single-variable analysis; this analysis should be routinely used in the evaluation of substrate performance.

Genetic variation for productive and adaptive characteristics has been reported in various *Eucalyptus* spp. [44–46]. There is information about variability in rooting and tolerance to salinity of *E. grandis* × *E. camaldulensis* hybrid [1,47]. However, there is a lack of information about the development of this eucalyptus clone in different substrate media. Sallesses et al. (2015) [13] evaluated different organic substrates in one eucalyptus clone. This work contributes to expanding the knowledge on the growth abilities of six Argentinean hybrid clones of eucalyptus in different substrates.

Rizzo et al. (2015) [24] have found that the compost formulated from initial mixtures with 40–60% poultry manure is alkaline. Although the authors did not find pH variations with the increase of poultry manure in the initial mixture, they found higher EC values in the treatment with 60% poultry manure. Generally, the compost derived from livestock manure had a lightly alkaline pH and a high salt content; therefore, it is necessary to amend these conditions before the use of such composts as substrate [30]. In this study, the initial pH values of the substrates were close to the optimum range.

The highest EC values of organic waste compost-based substrates were strongly influenced by the amount of poultry manure in the mixes, and ranged from 9.8 to 21.5 dS $m^{-1}$ when the proportions of poultry manure in the substrates increased from 0 to 41.7% [42].

Jayasinghe et al. (2010) [48] found that the components which most contribute to salinity are $Na^+$, $K^+$, ammonia, nitrate, and sulfate, which were found in greater amounts in compost-based substrates. The differences in the chemical and physicochemical properties of the organic wastes and in their proportions in the mixes markedly influenced the available contents of nutrients in the compost-based substrates [28].

In general, the organic waste compost-based substrates had EC values above the threshold limit level and high contents of available P, $K^+$, $NH_4^+$-N, and $Na^+$, while the commercial substrates had adequate levels of B for plants and presented lower concentrations of $NH_4^+$-N, P, $K^+$, and micronutrients. The compost-based substrate with 33.4% of poultry manure and 25% of sewage sludge in its composition approaches the commercial substrate in terms of eucalyptus biomass production among the organic wastes compost-based substrates. Salt leaching during irrigation could have mitigated the possible negative effects of high EC, which is an important attribute to consider when young plants are cultivated. However, salts can be managed to adequate levels [25]. Even the propagation of seedlings obtained from vegetative parts of eucalyptus clones instead of seeds can have their growth impaired by high salt concentrations, even if the plants do not exhibit any deleterious effects of salinity.

According to Marrón (2015) [49] the presence of heavy metals in the soil following the spreading of sewage sludge, wastewater, compost, manure, or ash has been shown by numerous studies, mainly in the upper soil horizons. He found that all studies were unanimous on the fact that the levels found were under the limit of the regulatory pollution thresholds (thresholds of the different countries where the experiments were carried out), did not present a risk for the environment or health, and were not leached to deep soil horizons or into groundwater. These low risks identified in short-rotation plantations were linked to the fact that willows, poplars, and eucalyptus were able to efficiently extract most soluble/exchangeable metals (Cd, Ni, and Zn) present in the sludge or manure as well as Cr, Pb, Hg, and Cu to a lesser extent.

In general, moderate nitrogen deficiency can improve rooting, although at high levels more energy is required for vegetative growth and especially for leaf expansion [50]. Consequently, carbohydrates are not stored at suitable levels, reducing the C:N ratio. On the other hand, extreme nitrogen deficiency can reduce rooting, as it is necessary for nucleic acid and protein synthesis. Based on the considerations of Hartman and Kester (1983) [51] and of Haissig (1973, 1986) [50,52], a general model for leaf nutrient concentration can be established. The mother plant should be well nourished with macronutrients such as P, K, Ca, and Mg, and should be moderately deficient in nitrogen.

Organic waste compost-based substrates present the highest available contents of $NH_4^+$-N and $NO_3^-$-N, which is possibly associated with the higher proportions of poultry and cattle manure, coffee husk, and sewage sludge in the mixes. In comparison to the commercial substrates, the levels of $NH_4^+$-N in the organic waste compost-based substrates are higher, and in most cases above the ideal range recommended by Abad et al. (2001) [10]. The clones (*E. grandis* × *E. urophylla*) preferentially absorbed $NH_4^+$-N in relation to $NO_3^-$-N. However, as verified by Atiyeh et al. (2000) [53], the high content of $NH_4^+$ in poultry manure could affect plant roots, which may cause detrimental effect in the plant growth.

Studies on the effects of various types of compost on forest production are quite disparate and highly variable. Composted manure did not increase root development or water status in poplars, but increased soil water retention properties. This type of compost caused growth gain. Compost of municipal solid residues stimulated the growth of poplar and eucalyptus. Compost or mulch of Southern fern moderately stimulated the growth of eucalyptus seedlings (+8 to 20%) [49]. In most cases, the spreading of sludge, pretreated wastewater, manure, compost, and industrial effluents actually caused an increase in organic carbon, organic matter, and nitrogen in the soil and increased the productivity of willows, poplars and eucalyptus, filling the deficiencies in N and K of the trees.

In this study, the highest values of eucalyptus growth were obtained with S3, the substrate made with compost derived from corn bare cobs and poultry manure. This might have happened because growth is a function of total pore space [54]. A substrate with low bulk density is economically favorable because it significantly improves the operational capacity of the growing medium, decreasing the cost of transport and manipulation of the materials [55].

Several authors agree with the results found in this study. Oliveira Junior et al. (2011) [56] concluded that the presence of livestock manure in the formulation of substrates gives rise to benefits such as an increase in the supply of nutrients, a reduction in costs in the production of seedlings, and a decrease in the dependence on forestry in relation to commercial substrates. On the other hand, Santos et al. (2000) [57] highlighted that it is difficult to find a material that meets all the requirements of the species that are cultivated. However, Da Silva et al. (2012) [58] obtained the best results for the production of clonal seedlings of *E. urophylla* × *E. grandis* hybrids using coconut husk fiber and carbonized rice husk as pure components of the substrate, indicating that it is possible to find materials that, in a pure state, can be used as substrates. In this sense, Sallesses et al. (2015) [13] obtained a better agronomic response in hybrid clones of *E. grandis* × *E. camaldulensis* with a substrate based on composted poultry manure (100%). Although this compost had limitations of use due to the high electrical conductivity, limited thermophilic phase, and a certain degree of toxicity [59], when used as a substrate for eucalyptus seedlings they obtained the best results. In this sense, composts of livestock origin, despite the limitations on their use due to their physicochemical properties, could be used as substrates in forestry.

These alternative substrates, when well characterized and corrected by suitable mixtures, make it possible to produce plants with better quality more rapidly and avoid over-exploitation of natural peatlands [60]. The use of those substrates contributes to resolving the problem of local waste generation (e.g., production waste and poultry manure). Alternative substrates must be increasingly used in order to include horticulture in sustainable agricultural systems [7].

An alternative to improve the agronomic performance of organic wastes composted-based substrates in order to reach the performance presented by commercial substrates would be to increase the addition rate of plant residues rather than animal waste, thereby decreasing EC and nutrients contents. The combination of peat with organic waste is another strategy that can reduce high salinity, heterogeneity, and/or high content of contaminants and nutrients in substrates [61].

Marrón (2015) [49] attempted to compile the results of studies about the effects of land applications of various kinds of residues (principally organic), including sewage and industrial sludge, wastewater, manure, compost, ashes, biochar, and landfill leachates in plantations, under natural conditions, or under controlled pot conditions of the three species mostly used for short-rotation coppice (SRC), namely, willow, poplar, and eucalyptus. Agronomic (at plant and soil levels) and environmental (soil and water contamination) effects were reviewed. The authors found that only 8% of the 288 references were from South America. In contrast to agriculture, where manure is primarily used, in short rotation coppice is used less than one third of the time. In the present study, we found clone preference and synergism for specific substrates. Zaller (2007) [62] highlighted differences of vermicompost effects between crop varieties and provided recommendations on the optimum proportion of vermicompost amendment to horticultural potting substrates.

These results suggest that further research is necessary on the final use of these composts in agricultural applications. This is a strategic way of solving a specific waste problem by recycling it and turning it into a valuable component for a cultivation substrate for specific primary productions. These results suggest that the interactions among substrates and clones are worthy of further investigation.

## 5. Conclusions

The studied substrates showed higher survival than that of the control. *E. grandis* × *E. camaldulensis* hybrid propagation could be carried out in alternative substrates formulated from poultry manure compost. The physicochemical characteristics of substrates composed of pine bark and poultry manure compost provide adequate conditions for the growth of eucalyptus. In this sense, the composting treatment of poultry manure and its use in forest nurseries contributes to the circular economy approach by minimizing negative effects on the environment.

GC8 was the genotype most sensitive to the use of different substrates, showing significant differences in several vegetal growth parameters (roots/callus ratio, radicular dry weight and cutting dry weight). This clone might be a good option to evaluate compost-based substrates for forestry applications.

**Author Contributions:** Conceptualization, P.F.R., D.K. and A.S.; methodology, P.F.R. and L.F.S.; software, V.D.T.; formal analysis, V.D.T.; investigation, P.F.R. and D.K.; resources, P.S.P.; data curation, V.D.T. and M.C.S.; writing—original draft preparation, P.F.R., V.D.T. and M.C.S.; writing—review and editing, P.F.R. and A.S.; visualization, L.F.S., N.I.R. and J.P.D.; supervision, L.F.S., N.I.R. and J.P.D.; project administration, P.F.R.; funding acquisition, P.F.R. and A.S. All authors have read and agreed to the published version of the manuscript.

**Funding:** This research was funded by PNNAT 1128042 and 2019-E2-I518-002 from INTA, Argentina.

**Acknowledgments:** Pedro F. Rizzo would like to thank INTA for the PhD research scholarship awarded.

**Conflicts of Interest:** The authors declare no conflict of interest.

## References

1. Ayala, P.; Surenciski, M.; Harrand, L.; Luna, C. Capacidad de enraizamiento de clones híbridos de Eucalyptus del Instituto Nacional de Tecnología Agropecuaria, Argentina. *Temas Agrar.* **2020**, *25*, 66–76. [CrossRef]
2. Cerasoli, S.; Caldeira, M.C.; Pereira, J.S.; Caudullo, G.; De Rigo, D. Eucalyptus globulus and other eucalypts in Europe: Distribution, habitat, usage and threats. In *European Atlas of Forest Tree Species*; San-Miguel-Ayanz, J., de Rigo, D., Caudullo, G., Durrant, T.H., Mauri, A., Eds.; Publishing Office of the EU: Luxembourg, 2016.
3. Eldridge, K.G.; Davidson, J.; Harwood, C.E.; van Wyk, G. *Eucalypt Domestication and Breeding*; Clarendon Press: Oxford, UK, 1994.
4. Marcó, M.; Harrand, L. Valor Potencial De Los Eucaliptos Colorados en combinaciones Hibridas. In Proceedings of the Jornada Sobre Potencialidad Foresto-Industrial del Eucalipto, Santiago del Estero, Argentina, 16–17 June 2005.
5. Abad, M.; Noguera, P.; Carrión, C. Los sustratos en los cultivos sin suelo. In *Tratado de Cultivo sin Suelo*; Mundi-Prensa: Madrid, Spain, 2004; pp. 113–158.
6. Ansorena, M.J. *Sustratos Propiedades y Caracterización*; Mundi-Prensa: Madrid, Spain, 1994; ISBN 84-7114-481-6.
7. Guérin, V.; Lemaire, F.; Marfa, O.; Caceres, R.; Giuffrida, F. Growth of Viburbum tinus in peat-bases and peat-subtitute growing media. *Sci. Hortic.* **2001**, *89*, 129–142. [CrossRef]
8. Jiménez, R.; Caballero, M.R. *El Cultivo Industrial de Plantas en Maceta*; Ediciones de Horticultura, S.L.: Reus, Spain, 1990.

9. Mañas, P.; Castro, E.; de las Heras, J. Quality of maritime pine (*Pinus pinaster* Ait.) seedlings using waste materials as nursery growing media. *New For.* **2009**, *37*, 295–311. [CrossRef]
10. Abad, M.; Noguera, P.; Burés, S. National inventory of organic wastes for use as growing media for ornamental potted plant production: Case study in Spain. *Bioresour. Technol.* **2001**, *77*, 197–200. [CrossRef]
11. Struve, D.K. Effect of copper-treated containers on transplant survival and regrowth of four three species. *J. Environ. Hortic.* **1993**, *11*, 196–199. [CrossRef]
12. García, C.; Alcántar, O.G.; Cabrera, R.I.; Gavi, F.; Volke, V. Evaluación de sustratos para la producción de *Epipremnum aureum* y *Spathiphyllum wallisii* cultivadas en maceta. *Terra* **2001**, *19*, 249–258.
13. Sallesses, L.F.; Rizzo, P.F.; Riera, N.I.; Della Torre, V.; Crespo, D.E.C.; Pathauer, P.S. Efecto de compost de guano avícola en la producción de clones híbridos de *Eucalyptus grandis* × *Eucalyptus camaldulensis*. *Cienc. Suelo* **2015**, *33*, 221–228.
14. Abad, M.; Noguera, P.; Noguera, V.; Roig, A.; Cegarra, J.; Paredes, C. Reciclado de residuos orgánicos y su aprovechamiento como sustratos de cultivo. *Acta Hortic.* **1997**, *19*, 92–109.
15. Maldonado-Benítez, R.K.; Aldrete, A.; Lopez-Upton, J.; Vaquera-Huerta, H.; Cetina-Alcalá, V.M. Producción de *Pinus greggii* Engelm. en mezclas de sustrato con hidrogel y riego en vivero. *Agrociencia* **2011**, *45*, 389–398.
16. Hernández-Apaolaza, L.; Gasco, A.; Gasco, J.; Guerrero, F. Reuse of waste materials as growing media for ornamental plants. *Bioresour. Technol.* **1983**, *96*, 125–131. [CrossRef]
17. Barbaro, L.; Karlanian, M.; Rizzo, P.F.; Riera, N. Caracterización de diferentes compost para su uso como componente de sustratos. *Chil. J. Agric. Anim. Sci.* **2019**, *35*, 126–136. [CrossRef]
18. Ko, H.J.; Kim, K.Y.; Kim, H.T.; Kim, C.N.; Umeda, M. Evaluation of maturity parameters and heavy metal contents in compost made from animal manure. *Waste Manag.* **2008**, *28*, 813–820. [CrossRef] [PubMed]
19. Rizzo, P.F.; Bres, P.A.; Young, B.J.; Zubillaga, M.S.; Riera, N.I.; Beily, M.E.; Argüello, A.; Crespo, D.E.C.; Sanchéz, A.; Komilis, D. Temporal variation of physico-chemical, microbiological, and parasitological properties of poultry manure from two egg production systems. *J. Mater. Cycles Waste Manag.* **2020**, *22*, 1140–1151. [CrossRef]
20. Pin Viso, N.D.; Rizzo, P.F.; Young, B.J.; Gabioud, E.; Bres, P.; Riera, N.I.; Merino, L.; Farber, M.; Crespo, D.C. The Use of Raw Poultry Waste as Soil Amendment Under Field Conditions Caused a Loss of Bacterial Genetic Diversity Together with an Increment of Eutrophic Risk and Phytotoxic Effects. *Microb. Ecol.* **2022**, *85*, 1–14. [CrossRef]
21. Kelleher, B.P.; Leahy, J.J.; Henthan, A.M.; O'Dwyer, T.F.; Sutton, D.; Leahy, M.J. Advances in poultry litter disposal technology—A review. *Bioresour. Technol.* **2002**, *83*, 27–36. [CrossRef]
22. Turan, N.G. The effects of natural zeolite on salinity level of poultry litter compost. *Bioresour. Technol.* **2008**, *99*, 2097–2101. [CrossRef] [PubMed]
23. Rizzo, P.F.; Young, B.J.; Viso, N.P.; Carbajal, J.; Martínez, L.E.; Riera, N.I.; Bres, P.A.; Beily, M.E.; Barbaro, L.; Farber, M.; et al. Integral approach for the evaluation of poultry manure, compost, and digestate: Amendment characterization, mineralization, and effects on soil and intensive crops. *Waste Manag.* **2022**, *139*, 124–135. [CrossRef]
24. Alegre, M.; Kaufmann, I.I.; Llorente, R.T.; Ferrere, P.; Butti, M.; Messa, J.C.; Contreras, C.I. *Análisis Espacial del Balance Energético Derivado de Biomasa: Metodología WISDOM Provincia de Buenos Aires*, 1st ed.; Organización de las Naciones Unidas para la Alimentación y la Agricultura-FAO: Buenos Aires, Argentina, 2018.
25. Rizzo, P.; Della Torre, V.; Riera, N.; Crespo, D.; Barrena, R.; Sánchez, A. Co-composting of poultry manure with other agricultural wastes: Process performance and compost horticultural use. *J. Mater. Cycles Waste Manag.* **2015**, *17*, 42–50. [CrossRef]
26. Hernandez-Zarate, L.; Aldrete, A.; Ordaz-Chaparro, V.M.; López-Upton, J.; López-López, M.A. Nursery growth of *Pinus montezumae* Lamb influenced by different substrate mixtures. *Agrociencia* **2014**, *48*, 627–637.
27. Marfà, O.; Tort, J.M.; Olivella, C.; Caceres, R.; Martinez, F.X. Cattle manure compost as substrate. II. Conditioning and formulation of growing media for pot plants and bag culture. International Symposium on composting and use of composted materials. In Proceedings of the International Society for Horticultural Science (ISHS), Leiden, The Netherlands, 1 July 1998.
28. Ao, Y.; Sun, M.; Li, Y. Effect of organic substrates on available elemental contents in nutrient solution. *Bioresour. Technol.* **2008**, *99*, 5006–5010. [CrossRef]
29. Valenzuela, O.R.; Gallardo, C.S.; Carponi, M.S.; Aranguren, M.E.; Tabares, H.R.; Barrera, M.C. Manejo de las propiedades físicas en sustratos regionales para el cultivo de plantas en contenedores. *Cienc. Docencia Tecnol.* **2014**, *4*, 1–19.
30. Waste and Resources Action Program (WRAP). Guidelines for the Specification of Quality Compost for Use in Growing Media. 2011. Available online: www.wrap.org.uk (accessed on 17 October 2022).
31. SCyMA, SENASA. Resolution 1/19-Regulatory Framework for the Production, Registration and Application of Compost. Secretariat of Enviromental Control and Monitoring, Argentina. 2019. Available online: www.argentina.gob.ar (accessed on 17 October 2022). (In Spanish).
32. United States Department of Agriculture (USDA); United States Composting Council (USCC). *Test Methods for the Examination of Composting and Compost*, 1st ed.; Edaphos International: Houston, TX, USA, 2001.
33. Resolución SENASA 264. Manual para el Registro de Fertilizantes, Enmiendas, Sustratos, Acondicionadores, Protectores y Materias Primas en la República Argentina. 2011. Available online: www.senasa.gob.ar (accessed on 16 October 2022). (In Spanish)..
34. López, J. Variación genética de propiedades físicas organolépticas de importancia para usos solidos de alto valor de clones híbridos de Eucalyptus en Entre Ríos. *Actas XXXI Jorn. For. Entre Ríos* **2017**, *1*, 67–76.

35. Alliani, R.C. Plan de Mejoramiento en la Región Pampeana. In *Jornada sobre Eucaliptos para la Región Pampeana*; Actas Centro de Investigaciones y Experiencias Forestales (CIEF): Buenos Aires, Argentina, 1990; pp. 23–28.
36. Harrand, L.; Schenone, R. Obtención de individuos híbridos de Eucalyptus a través de la realización de cruzamientos controlados. In Proceedings of the XVII Jornadas Forestales de Entre Ríos, Concordia, Argentina, 13–14 October 2002; p. 10.
37. Harrand, L.; Oberschelp, J.; Salto, C.; Marcó, M.; López, J.A. Ensayos clonales de especies e híbridos de Eucalyptus. *Domest. Mejor. Especies For.* **2016**, *1*, 61–62.
38. INTA. *Compendium of Analytical Methods for the Characterization of Agricultural and Agro-Industrial Waste, Compost and Effluents*, 1st ed.; Ediciones INTA: Buenos Aires, Argentina, 2021. (In Spanish)
39. Di Rienzo, J.A.; Casanoves, F.; Balzarini, M.G.; Gonzalez, L.; Tablada, M.; Robledo, C.W. *InfoStat*; Versión, 2011; Grupo InfoStat, FCA; Universidad Nacional de Córdoba: Córdoba, Argentina, 2011; Available online: http://www.infostat.com.ar (accessed on 10 October 2022).
40. Barbaro, L.A.; Karlanian, M.A. Uso del compost en la formulación de sustratos para plantas. In *Compostaje en la Argentina: Experiencias de producción, calidad y uso*; Mazzarino, M.J., Satti, P., Eds.; Universidad Nacional de Rio Negro y Orientación Gráfica Editora: Buenos Aires, Argentina, 2012; ISBN 9789879260937.
41. Woodward, A.J.; Bennett, I.J.; Pusswonge, S. The effect of nitrogen source and concentration, medium pH and buffering on in vitro shoot growth and rooting in *Eucalyptus marginata*. *Sci. Hortic.* **2006**, *110*, 208–213. [CrossRef]
42. Higashikawa, F.S.; Silva, C.A.; Nunes, C.A.; Bettiol, W.; Guerreiro, M.C. Physico-Chemical Evaluation of Organic Wastes Compost-Based Substrates for Eucalyptus Seedlings Growth. *Commun. Soil Sci. Plant Anal.* **2016**, *47*, 581–592. [CrossRef]
43. Rinaldi, S.; De Lucia, B.; Salvati, L.; Rea, E. Understanding complexity in the response of ornamental rosemary to different substrates: A multivariate analysis. *Sci. Hortic.* **2014**, *176*, 218–224. [CrossRef]
44. Cappa, E.P.; Pathauer, P.S.; Lopez, G.A. Provenance variation and genetic parameters of *Eucalyptus viminalis* in Argentina. *Tree Genet. Genomes* **2010**, *6*, 981–994. [CrossRef]
45. Dillon, S.; McEvoy, R.; Baldwin, D.S.; Rees, G.N.; Parsons, Y.; Southerton, S. Characterisation of adaptive genetic diversity in environmentally contrasted populations of *Eucalyptus camaldulensis* Dehnh. (river red gum). *PLoS ONE* **2014**, *9*, e103515. [CrossRef]
46. Okun, D.O.; Kenya, E.U.; Oballa, P.O.; Odee, D.W.; Muluvi, G.M. Analysis of genetic diversity in *Eucalyptus grandis* (Hill ex Maiden) seed sources using inter simple sequence repeats (ISSR) molecular markers. *Afr. J. Biotechnol.* **2008**, 7.
47. Lopez, J.A.; Harrand, L.; Marco, M.A.; Lopez, A.J. Variación genética de clones híbridos de Eucalyptus. *Quebracho* **2016**, *24*, 5–6.
48. Jayasinghe, G.Y.; Tokashiki, Y.; Kitou, M. Use of synthetic soil aggregates as a containerized growth medium component to substitute peat in the ornamental plant production. *Arch. Agron. Soil Sci.* **2010**, *56*, 183–199. [CrossRef]
49. Marrón, N. Agronomic and environmental effects of land application of residues in short-rotation tree plantations: A literature review. *Biomass Bioenergy* **2015**, *81*, 378–400. [CrossRef]
50. Haissig, B.E. Metabolic processes in adventitious rooting of cuttings. In *New Root Formation in Plants and Cuttings*; Jackson, M.B., Ed.; Martinus Nijhof Publishers: Dordrecht, The Netherlands, 1986.
51. Hartman, H.T.; Kester, D.E. *Plant Propagation Principles and Practices*; Prentice-Hall: Hoboken, NJ, USA, 1983.
52. Haissig, B.E. Metabolism during adventitious root primordium initiation and development. *N. Z. J. For. Sci.* **1973**, *4*, 324–337.
53. Atiyeh, R.M.; Subler, S.; Edwards, C.A.; Bachman, G.; Metzger, J.D.; Shuster, W. Effects of vermicomposts and composts on plant growth in horticultural container media and soil. *Pedobiologia* **2000**, *44*, 579–590. [CrossRef]
54. Chong, C.; Cline, R.A.; Rinker, D.L. Bark and peat amended spent mushroom compost for containerized culture of shrubs. *Hortic. Sci.* **1994**, *29*, 781–784. [CrossRef]
55. Abad, M.; Martínez, P.F.; Martínez, M.D.; Martínez, J. Evaluación agronómica de los sustratos de cultivo. *Acta Hortic.* **1993**, *11*, 141–154.
56. Oliveira Júnior, O.A.D.; Cairo, P.A.R.; Novaes, A.B.D. Características morfofisiológicas associadas à qualidade de mudas de *Eucalyptus urophylla* produzidas em diferentes substratos. *Rev. Arvore* **2011**, *35*, 1173–1180. [CrossRef]
57. Santos, C.B.; Longhi, S.J.; Hoppe, J.M.; Moscovich, F.A. Efeito do volume de tubetes e tipo de substratos na qualidade de mudas de *Cryptomeria japonica* (LF) D. Don. *Ciênc. Florest.* **2000**, *10*, 1–15. [CrossRef]
58. Da Silva, R.B.; Simões, D.; Da Silva, M.R. Qualidade de mudas clonais de *Eucalyptus urophylla* × *E. grandis* em função do substrato. *Rev. Bras. Eng. Agric. Ambient.* **2012**, *16*, 297–302. [CrossRef]
59. Young, B.J.; Rizzo, P.F.; Riera, N.I.; Torre, V.D.; López, V.A.; Molina, C.D.; Fernández, F.E.; Crespo, D.C.; Barrena, R.; Komilis, D.; et al. Development of phytotoxicity indexes and their correlation with ecotoxicological, stability and physicochemical parameters during passive composting of poultry manure. *Waste Manag.* **2016**, *54*, 101–109. [CrossRef]
60. Calkins, J.B.; Jarvis, B.R.; Swanson, B.T. Compost and rubber tire chips as peat substitutes in nursery container media: Growth effects. *J. Environ. Hortic.* **1997**, *15*, 88–94. [CrossRef]
61. Raviv, M.; Wallach, R.; Blom, T.J. The effect of physical properties of soilless media on plant performance-A review. In Proceedings of the International Symposium on Growing Media and Hydroponics, Alnarp, Sweden, 1 February 2004.
62. Zaller, J.G. Vermicompost as a substitute for peat in potting media: Effects on germination, biomass allocation, yields and fruit quality of three tomato varieties. *Sci. Hortic.* **2007**, *112*, 191–199. [CrossRef]

processes

MDPI

Article

# Treatment of Sewage Sludge Compost Leachates on a Green Waste Biopile: A Case Study for an On-Site Application

**Chaher Ibrahim Irka [1,*], Pascale Prudent [1], Frédéric Théraulaz [1], Anne-Marie Farnet Da Silva [2], Laurence Asia [1], Didier Gori [1], Laurent Vassalo [1], Amandine Durand [1], Carine Demelas [1], Patrick Höhener [1] and Pascal Wong-Wah-Chung [1]**

1 Aix Marseille Univ, CNRS, LCE, UMR 7376, 13331 Marseille, France; pascale.prudent@univ-amu.fr (P.P.); frederic.theraulaz@univ-amu.fr (F.T.); laurence.asia@univ-amu.fr (L.A.); didier.gori@univ-amu.fr (D.G.); laurent.vassalo@univ-amu.fr (L.V.); amandine.durand@univ-amu.fr (A.D.); carine.demelas@univ-amu.fr (C.D.); patrick.hohener@univ-amu.fr (P.H.); pascal.wong-wah-chung@univ-amu.fr (P.W.-W.-C.)

2 Aix Marseille Univ, Avignon Université, CNRS, IRD, IMBE, UMR 7263, 13331 Marseille, France; a-m.farnet@univ-amu.fr

* Correspondence: chaher.ibrahim-irka@etu.univ-amu.fr

**Abstract:** This work proposes a suitable treatment for the leachates from a sewage sludge composting process using a specific windrow (biopile). The biopile's evolution and organic content degradation were followed for 2 months with regular leachate spraying to assess the physico-chemical and biological impacts, and determine the risk of enrichment with certain monitored pollutants. The final objective was the valorization of the biopile substrates in the composting process, while respecting the quality standards of use in a circular economy way. Classical physico-chemical parameters (pH, conductivity, dissolved organic carbon (DOC), total dissolved nitrogen (TDN), etc.) were measured in the leachates and in the water-extractable and dry-solid fractions of the biopile, and the catabolic evolution of the micro-organisms (diversity and activities), as well as the enrichment with persistent organic pollutants (POPs) (prioritized PAHs (polycyclic aromatic hydrocarbons) and PCBs (polychlorinated biphenyls)), were determined. The results showed that the microbial populations that were already present in the biopile, and that are responsible for biodegradation, were not affected by leachate spraying. Even when the studied compost leachate was highly concentrated with ammonium nitrogen (10.4 gN $L^{-1}$ on average), it significantly decreased in the biopile after 2 weeks. A study on the evolution of the isotopic signature ($\delta^{15}$ N) confirmed the loss of leachate nitrogen in its ammoniacal form. The bio-physico-chemical characteristics of the biopile at the end of the experiment were similar to those before the first spraying with leachate. Moreover, no significant enrichment with contaminants (metal trace elements, volatile fatty acids, or persistent organic pollutants) was observed. The results show that it would be possible for composting platforms to implement this inexpensive and sustainable process for the treatment of leachates.

**Keywords:** compost; leachate treatment; reuse on site; sewage sludge; green waste

**Citation:** Ibrahim Irka, C.; Prudent, P.; Théraulaz, F.; Farnet Da Silva, A.-M.; Asia, L.; Gori, D.; Vassalo, L.; Durand, A.; Demelas, C.; Höhener, P.; et al. Treatment of Sewage Sludge Compost Leachates on a Green Waste Biopile: A Case Study for an On-Site Application. *Processes* **2022**, *10*, 1196. https://doi.org/10.3390/pr10061196

Academic Editor: Blaž Likozar

Received: 31 May 2022
Accepted: 13 June 2022
Published: 15 June 2022



## 1. Introduction

The European Union's waste management policy, which has been implemented for more than a decade [1], continues to influence the actions of the EU member states, with increasingly ambitious targets being set over the years. One of the main objectives is to move towards a sustainable development model known as the circular economy [2]. In France, the National Waste Prevention Program (Programme National de Prévention des Déchets—PNPD) guides the strategies for the public policy of waste prevention and plans to reduce household and similar waste by 15% for the 2021–2027 edition [3]. In the reduction of waste quantities and the circular economy model, the recovery of non-avoidable wastes is a priority, and material recovery is prioritized before energy recovery. Among the material recovery routes, composting and methanization are often considered for urban organic waste (bio-waste, green waste, sludge from sewage treatment plants, etc.).

The composting process is an aerobic transformation of fermentable materials under controlled conditions. Two phases follow one another during composting: a bio-oxidation and hygienization phase with a rise in temperature, and a maturation phase with a drop in temperature, which makes it possible to obtain a stable material rich in humic compounds. Stabilized compost is mainly used as an organic amendment [4] and also in bioremediation [5]. This technique thus enables a large proportion of fermentable organic waste to be valorized via material recovery.

Sewage sludge, the final residue of urban wastewater treatment, is well-suited to this type of recovery. The MIATE (materials of agronomic interest derived from water treatment) compost in France, which is generally obtained by a joint treatment with green waste, can be used in agriculture according to the NFU 44-095 quality standard of AFNOR (Association Française de Normalisation) [6]. This standard specifies the thresholds for certain physico-chemical and biological parameters in order to guarantee the quality of the compost produced for an organic soil amendment. In particular, the total phosphorus in $P_2O_5$ must be less than 3% by raw weight (RW), the dry weight (DW) expressed as raw weight must be greater than or equal to 50%, the sum of seven monitored PCBs (polychlorinated biphenyls) must not exceed 0.8 mg kg$^{-1}$ DW, and limit values are given for some PAHs (polycyclic aromatic hydrocarbons) (4 mg kg$^{-1}$ DW for fluoranthene, 2.5 mg kg$^{-1}$ DW for benzo(b)fluoranthene, and 1.5 mg kg$^{-1}$ DW for benzo(b)pyrene) and for some trace metal elements (TMEs) (300 mg kg$^{-1}$ DW for Cu, 180 mg kg$^{-1}$ DW for Pb, and 3 mg kg$^{-1}$ DW for Cd).

Regardless of the method used (mixture of several substrates or not), the composting process leads to the production of leachates, or compost juices, resulting from the percolation of water (process and metabolic water) in the materials treated mainly during the bio-oxidation phase. Several organic (dissolved and particulate organic matter, volatile fatty acids) and inorganic (ammonium, nitrates, phosphates) compounds are present in leachates, as well as pathogens (*Salmonella* spp., *Listeria monocytogenes*, etc.) and, potentially, inorganic and organic pollutants (PCBs, cadmium, copper, lead, etc.) [7]. Leachate production amounts vary between 5 and 170 L t$^{-1}$ of waste, according to the literature [8]. In France, these liquids have to be treated and/or eliminated, according to Article 22 of the Decree of 22 April 2008, governing the Classified Installations for the Protection of the Environment (ICPE) [9]. The choice of draining leachates by returning them to a wastewater treatment plant (WWTP), which is often the least technically restrictive option, generally represents a high cost, and regulatory constraints compromise this solution. Among the most widely used alternative techniques to treat leachates are reverse osmosis [10], evaporation [11], and electrochemical oxidation [12]. These technologies require a heavy investment for the acquisition and the daily operating cost. Other more rustic techniques are used to treat leachates, such as reed filters [13,14] and agro-industrial materials [15]. Concerning the reed filter, this method requires a substantial layout to support the continuous flow of effluent and a highly optimized sizing in order to limit the saturation of the system, which is highly conducive to odor pollution and problematic for sites in urban areas. For the other rustic methods, their application on an industrial scale deserves further investigation.

The aim of this research was to evaluate a low-cost alternative procedure to treat, and reuse on site, the compost leachates produced during the bio-oxidation phase of composting, in accordance with a circular economy approach and its economic and environmental issues. For this purpose, we worked on a platform producing MIATE compost, which generates leachates with a high organic matter content.

The treatment process was based on the spraying of leachates at regular time intervals on selected green waste biopiles in order to favor the liquid retention and biodegradation of the organic load of the leachates coming from the sewage sludge bio-oxidation step of composting (mixed with green wastes). The potential degradation of a specific biopile and the assessment of the potential risk of its enrichment with certain tracked pollutants or unwanted compounds were investigated through physico-chemical analysis, microbiological activity, and microbiological diversity studies. The obtained results could hypothetically validate the ultimate the valorization of biopile substrates in the composting process in a circular economy way, respecting the quality standards of use for MIATE composts.

## 2. Materials and Methods

### 2.1. Study Site: Setting up of the Biopile and Trial Progress

The study was performed at the SARL Biotechna composting platform (43°22′37.4″ N 5°11′27.8″ E) located in Ensues-la-Redonne, France. Biotechna has been an ICPE site since 2004 and treats up to 30,000 t of green waste and 30,000 t of WWTP sludge each year to produce MIATE compost according to the French standard (NFU 44-095). During the industrial composting process, between 2000 and 3000 $m^3$ of leachates (derived from the bio-oxidation step) are produced each year. These leachates are stored in a watertight tank before their treatment in WWTPs.

The biopiles were composed of a mixture of 80% crushed green waste (size: 80–110 mm) and 20% green waste compost (size: 20 mm) favorable to leachate retention and treatment conditions. This composition was selected after optimization pre-studies, considering the materials that are available on site in a self-centered circular economy (results not presented here). These pre-studies consisted of various mesocosm tests with different mixtures of crushed green waste (with or without sieving—20 mm) and green waste compost, through which the composition was optimized to guarantee the retention capacity and biodegradation of the leachate (a majority of crushed green waste allowed structuration of the biopile and a good aeration to favor biological activity). The dimensions of the experimental biopiles were 5 m long, 4 m wide, and 2.3 m high. The total surface area covered was 20 $m^2$, for a total volume of 46 $m^3$. Watering was carried out on a surface of 12 $m^2$ at the top of the biopile to avoid rapid percolation in the peripheral areas and the loss of leachates. A sprinkler system with 6 sprinklers (Ref. Sprinkler head—Pendent K80 DN 15 $\frac{1}{2}$ Male—68°—TY325) overhung the windrow and sprayed 1.5 $m^3$ (flow rate = 37.5 L $min^{-1}$) during each watering (the equivalent of 5.4% of the sprayed volume of the biopile). Furthermore, these watering conditions ensured an optimal retention capacity and avoided percolation at the bottom of the biopile and potential olfactory disturbances due to bad odors. The experiment was undertaken over 8 weeks, for a total of 4 waterings, to examine the potential for biopile enrichment with tracked pollutants or unwanted compounds. A watering frequency of 2 weeks allowed for the observation of the potentially significant evolutions of the monitored analytical parameters. The different steps and sampling campaigns carried out during the trial are presented in the Supplementary Materials, Table S1. For each watering step, 3 sampling campaigns were defined: sampling after the supply was completed ("after supply"), sampling one week after watering ("after 1 week"), and sampling two weeks after watering (and before the next watering) ("after 2 weeks"). During each watering step, samples were systematically collected before and after watering. A control biopile was monitored during the experiment, with the same characteristics (composition and size) and conditions of spraying as the test biopile, but with clean water instead of compost leachate.

The analyses of the liquid fractions (water extracts and leachates) were carried out on the quantification of dissolved organic carbon, total dissolved nitrogen, phosphate, nitrate, volatile fatty acids, and the monitoring of some spectroscopic indices, such as specific ultraviolet absorbance at 254 nm ($SUVA_{254}$) and the $E_2/E_3$ index (ratio of Abs 250 nm/Abs 365 nm, with Abs meaning absorbance). For the solid phase, the total carbon, total nitrogen with nitrogen isotopy ($\delta^{15}$ N), total trace metal elements, and total phosphorus, as well as some PAHs and PCBs of interest, were monitored.

### 2.2. Sampling and Pretreatment of Solid Samples

During each sampling campaign, 3 solid samples were collected randomly from the windrow at a depth between 30 and 80 cm, in order to accommodate the sampling variability of such a heterogeneous substrate. Two were located oppositely and at the periphery of the sprayed area, and one was located in the center of this area. Each sample was sieved using a 50 mm mesh sieve before 3 successive quarterings to obtain a representative sample of about 1 kg, which was distributed into bags and stored at 4 °C for further analysis.

The dry-solid samples obtained after oven drying at 105 °C were grinded in 3 steps: 2 successive grindings in a Fristch mill with a sieve size of 2 mm followed by a sieve size of 0.25 mm, and a final grinding in an RM 200 mortar grinder to obtain fine particles for the analysis of the solids.

### 2.3. Sampling and Pretreatment of Liquid Samples

During each watering campaign, 6 L of leachate was sampled and stored in amber glass bottles in darkness at 4 °C prior to analysis.

The water extracts [16] obtained from fresh samples were carried out according to the following protocol: 6 g of fresh solid was mixed with 60 mL of ultra-pure water produced by a Millipore water system (Milli-Q® Direct, Darmstadt, Germany). The samples were shaken at 125× *rpm* for 2 h and then centrifuged at 8000× *rpm* for 15 min at 20–22 °C. Both the supernatants and the leachate samples were filtered through 0.45 µm PolyEtherSulfone filters before the analysis.

For each sampling campaign, the obtained analytical values corresponded to the average and confidence interval for the 3 collected samples ($n = 3$). The result for each compound (i) analyzed in the water extracts was expressed using the following Equation (1):

$$\text{Concentration (i)} = \frac{\text{C mg L}^{-1} * \text{V(L)}}{\text{Dry Weight (g)}} = \text{X mg g}^{-1}\text{ DW} \tag{1}$$

where the dry weight (DW) is the exact mass obtained after drying at 105 °C, C is the measured concentration of a physico-chemical value in solution, and V is the volume of the ultra-pure water used for the water extraction, which corresponds to 0.06 L.

### 2.4. Sample Treatment and Analysis

#### 2.4.1. Measurements of pH, Conductivity, Temperature and Humidity

The pH and conductivity were measured in the water extracts and leachate samples using multi 3420 portable probes provided by WTW.

Temperature measurements were obtained on site using a TCK mobile probe during the sampling campaigns in the biopile at a depth of 50 cm. To evaluate the humidity of the samples, approximately 200 g of fresh (raw) sample was placed in an oven at 105 °C. The drying was monitored for 72 h, and the weight was measured regularly until a stable dry weight was reached.

The moisture X was calculated according to the following Equation (2):

$$\text{Moisture (\%)} = \frac{\text{Raw weight (g)} - \text{Dry weight (g)}}{\text{Raw Weight (g)}} = \text{X} \tag{2}$$

#### 2.4.2. Water Extract (WE) Analysis

Dissolved Organic Carbon (DOC) and Total Dissolved Nitrogen (TDN) Analysis

The measurements of the DOC and TDN were carried out with a multi N/C® 2100 S [17]. This method allowed for the simultaneous determination of the carbon and nitrogen content in the water extract samples. Each sample (250 µL) was injected twice, and the average was reported as the DOC (mgC $L^{-1}$) or TDN (mgN $L^{-1}$). The detection limits were 3.19 mgC $L^{-1}$ for DOC and 1.16 mgN $L^{-1}$ for TDN.

Ammonium ($NH_4^+$) and Volatile Fatty Acid (VFA) Analysis—Fluorescence Microplate Reagents:

The reagents used in this work were: o-phthaldialdehyde (OPA), *N*-acetylcysteine (NAC), *N′*-(3-dimethylaminopropyl)-*N*-ethylcarbodiimide (EDC), *N*-(1-naphthyl)ethylenediamine (EDAN), 7-aza-1-hydroxybenzotriazole (HOAT), and absolute ethanol (EtOH).

- Ammonium

Ammonium was assayed using a microplate method with fluorescence detection (top configuration), following the formation of a fluorescent derivative (isoindole) by the reaction of the ammonium ions, OPA and NAC [18]. After preheating the microplate reader (Synergy™ HTX Multi-Detector Microplate Reader—BioteK) to 30 °C, 100 μL of the sample was dropped into the well; then, 20 μL of NAC (20 mM) was added, followed by 30 μL of OPA (13 mM). After shaking inside the reader (orbital mode with an amplitude of 3 mm), the fluorescence was measured at $\lambda_{exc}$ = 415 nm and $\lambda_{em}$ = 485 nm.

- Volatile Fatty Acids

The volatile fatty acids were analyzed in the same microplate reader with a fluorescence detection method [19]. The reader was preheated to 40 °C before depositing the following: 150 μL of HOAT (6 g $L^{-1}$) and EDAN (1 g $L^{-1}$) at a pH of 5; 30 μL of the sample; and 20 μL of EDC at 100 g $L^{-1}$ in EtOH. After first shaking for 5 min, 25 μL of $KH_2PO_4$ (250 mM) in NaOH (0.35 M) was added before a second shaking for 2 min. Finally, 20 μL of OPA (60 g $L^{-1}$) in EtOH was added before a third shaking for 5 min. The readings were taken at $\lambda_{ex}$ = 335 nm and $\lambda_{em}$ = 442 nm.

Anion Analysis—Nitrate ($NO_3^-$) and Phosphate ($PO_4^{3-}$)

The anions were analyzed using ion chromatography on an ICS-3000 HPLC system (Dionex) equipped with an AG11-HC guard column, an AS11-HC analytical column (4 × 250 mm), a Dionex CD-25 conductivity detector, and a 200 μL injection loop. The analyses were performed in isocratic mode (22.5 mM NaOH in helium-purged deionized water) at 30 °C, with a flow rate of 1.5 mL $min^{-1}$. An external electrochemical suppression system (ACRS 500 4 mm) was added to improve the signal-to-noise ratio [20]. The system control and data processing were performed using the Chromeleon7® software (ThermoFisher Scientific, Waltham, MA, USA). The quantification limits were 0.06 mgN $L^{-1}$ for $NO_3^-$ and 0.16 $mgP_2O_5$ $L^{-1}$ for $PO_4^{3-}$.

Monitoring of Spectroscopic Indices

The molecular properties of the dissolved organic material were monitored by UV-visible absorption. The spectra were performed on a Shimadzu UV-1800 spectrometer over a wavelength range of 200 to 1000 nm using a 10 mm quartz optical cell—3 mL volume. We calculated two spectroscopic indices to evaluate the evolution of the organic matter contained in the water extracts, representing WEOM (water-extractable organic matter) [16,21,22].

- The $E_2/E_3$ ratio was used to monitor the degree of aromaticity of the organic matter and the molecular size [23]. It was calculated as follows:

  $E_2/E_3$ = Abs 250 nm/Abs 365 nm
  Abs: Absorbance

- The $SUVA_{254}$ was measured to determine the abundance of aromatic carbons and provide a quantitative value of their content per unit concentration of dissolved organic carbon [24,25]. The $SUVA_{254}$ was obtained by dividing its absorbance at 254 nm by its DOC value (mgC $L^{-1}$) and the cell path length (m), and was expressed in L $mgC^{-1}$ $m^{-1}$.

### 2.4.3. Analysis of Solid Substrates

Total Carbon (TC) and Total Nitrogen (TN) Analysis

Elemental and isotopic analyses were performed using a coupled elemental analyzer (EA Flash HT Plus, Thermo Fisher Scientific, USA) with a mass spectrometer (IrMS Delta V Advantage, Thermo Fisher Scientific). Stable nitrogen analyses were performed via a combustion reactor (1000 °C) under a carrier gas flow (helium: 100 mL $min^{-1}$) with the sample sealed in a tin capsule and introduced via an automatic injector. The gases were separated on a chromatographic column before being quantified via the catharometric detector, and were then introduced into the mass spectrometer for isotopic analysis. The $\delta^{15}$ N ratio (15 N/14 N = $\delta^{15}$ N) analyses were expressed relative to atmospheric nitrogen [26–28].

Analysis of Total Phosphorus (TP) and Trace Metal Elements (TMEs)

Grinded solid samples were mineralized using a Start D microwave digester (Milestone, Brondby, Denmark). A total of 0.2 g of the finely ground sample (Section 2.2) was mineralized in 3 mL of Trace Metal Grade hydrochloric acid (Fisher Chemical Illkirch, Illkirch-Graffenstaden, France, HCl 35–38%) and 6 mL of Trace Metal Grade nitric acid (Fisher Chemical Illkirch, France, $HNO_3$ 67–69%). After recovering the sample in a 25 mL volumetric flask with ultrapure water, the sample was filtered through a 0.45 µm cellulose ester membrane filter.

The mineralized samples were analyzed using inductively coupled plasma optical emission spectrometry (ICP-OES) [29–31]. A JY 2000 Ultrace Sequential (Horiba Jobin Yvon, Edison, NJ, USA) was used, with a 2400 line $min^{-1}$ array, a monochromatic system, a cyclonic quartz spray chamber, and a radial plasma measurement system. Quality controls and accuracy were checked using two certified reference materials (CRM) of plant (CRM DC 73349) and soil (CRM DC 73323) with accuracies within 100 ± 10%.

Analysis of Persistent Organic Pollutants (POPs)

The solid samples for PAH and PCB analysis, stored at −80 °C, were freeze-dried before being ground to obtain particles of 0.25 mm. The extraction was performed in an ultrasonic bath for 15 min on 1 g of sample spiked with deuterated standards, which was placed in 40 mL of a solvent mixture of 1/1 hexane (n-Hexane ≥ 97%, HiPerSolv CHROMANORM® for HPLC) and dichloromethane (Dichloromethane ≥ 99.8%, HiPerSolv CHROMANORM® for HPLC). The supernatant was filtered through a 0.45 µm PTFE (polytetrafluoroethylene) filter. The extract was concentrated to a volume of 5 mL under a gentle stream of $N_2$ with Turbovap® II at 35 °C. The extract was finally purified on a Florisil SPE cartridge (1 g 6 mL) provided by Restek using a two-step elution with hexane (25 mL) and dichloromethane (5 mL).

A GC–MS method for PAH and PCB identification and quantification was modified from a previous study [32]. The GC–MS system was a Clarus 600/600 C from Perkin Elmer, equipped with a programmable temperature-vaporizing (PTV) injector programmed from 50 to 250 °C at 200 °C $min^{-1}$. This enabled the identification and quantification of 7 PCBs (PCB 28, PCB 52, PCB 101, PCB 118, PCB 138, PCB 153, and PCB 180) and 3 PAHs (fluoranthene, benzo(b)fluoranthene, and benzo(a)pyrene). However, benzo(b)fluoranthene and benzo(k)fluoranthene could not be separated with the analytical method and thus, the compound analyzed was benzo(b,k)fluoranthene. A Perkin Elmer DB-5 MS column (30 m × 0.25 mm × 0.25 µm) was used and 1 µL was injected in the splitless mode. A specific oven temperature program from 60 to 280 °C was developed to separate all compounds: 60 °C for 1 min, heated to 150 °C at 40 °C $min^{-1}$, then heated to 280 °C at 6 °C $min^{-1}$ and held for 20 min. Helium was used as the carrier gas at a constant flow rate of 1 mL $min^{-1}$. The ion source temperature was fixed at 250 °C and compound ionization was performed by electron impact ionization (70 eV). The quadrupole mass spectrometer operated with the simultaneous collection of a full scan and selected ion recording (selected ion full ion mode, or SIFI mode). The limits of detection (LOD) for the PCBs and PAHs ranged from 2.45 to 19.11 µg $kg^{-1}$ DW and 2.13 to 11.70 µg $kg^{-1}$ DW, respectively.

#### 2.4.4. Microbiological Analyses

The functional diversity of bacterial communities was studied using Ecoplate™ biolog microplates, which contained 96 wells corresponding to three repetitions each of 31 carbon sources and three wells containing water as controls [17,33]. After a respiratory process, the use of a carbon source led to the reduction of triphenyltetrazolium (TTC) to formazan, absorbing at 590 nm.

A fresh sample equivalent to 5 g of dry weight, stored for 24 h at 4 °C after collection, was added to 100 mL of 0.1% sodium pyrophosphate and stirred at 80× *rpm* for 2 h. An amount of 2 mL of the "S1" supernatant was placed in an Eppendorf tube, then centrifuged at 500× *rpm* for 10 min to obtain the "S2" supernatant. The organic matter was standardized by measuring the absorbance of the "S2" supernatant at 600 nm, diluted ten times with sterile 0.85% NaCl solution. When necessary, the "S2" supernatant was re-diluted with sterile 0.85% NaCl to obtain a final volume of 40 mL with an Abs = 0.03.

Two indices were calculated after measuring the absorbances after 48 h of incubation at 40 °C: The Shannon–Weaver index, H′, which provided information on the diversity of the bacterial communities in the sample [34], and the AWCD (average well color development) index, which provided information on the overall microbial catabolic activity present in the sample [35].

- The Shannon–Weaver index was calculated using the following formula:

$H' = -\sum Pi\ lnPi$, where $Pi = Abs(i)/\sum Pi$, where Abs(i) is the optical density for each positive well i and $\sum Pi$ is the sum of the color development of all positive wells.

- The AWCD index was calculated using the following formula:

$AWCD = \sum Abs(i)/31$, where Abs(i) is the optical density for each positive well (i), and 31 is the number of wells.

### *2.5. Data Clustering for Statistical Analysis*

The objectives were to study the potential abatement and fate of certain compounds and physico-chemical parameters of the leachate after several inputs into the biopile. As seen before, after each leachate addition, three sampling steps were investigated (Section 2.1). For each measured parameter, the average values calculated on all watering steps according to these three sampling steps led to statistical comparisons between these three steps, in order to evaluate the abatement over a period of 2 weeks after spraying the biopile with compost leachate. The first watering was discarded from the dataset for statistical processing because the obtained values corresponded to an anomaly. The non-parametric Kruskal–Wallis test with Dunn's procedure was applied to evaluate significant differences between the measured averages. The statistical analyses were performed using XLSTAT statistical and data analysis solutions (Paris, France; https://www.xlstat.com).

PHREEQC (https://www.usgs.gov/software/phreeqc-version-3, 2021) was used to model the speciation of ammonia nitrogen in the biopile.

The Ecoplate™ biolog values (AWCD values obtained for each well) were also analyzed by principal component analysis (PCA) to explain the variance between the different microcosms. For each treatment, the barycenter of the 3 replicates was calculated. The PCA approach, carried out on AZURAD® (http://www.azurad.fr/logiciel-plans-experiences.php), made it possible to project the quantitative variables (substrates consumed) and the individuals (sampling campaigns) measured from the biologs.

## 3. Results and Discussion

### *3.1. On-Site Characteristics of the Biopile*

The average temperatures were, respectively, 74.0 ± 4.6 °C (*n* = 8) and 75.0 ± 5.3 °C (*n* = 8) for the biopile watered with clean water (control biopile) and for the biopile watered with leachate (test biopile). The humidity was equal to 53.0 ± 3.4% on average for the control biopile and 48.0 ± 4.5% for the test biopile. These values corresponded to optimal conditions for microbial growth during the thermophilic phase [36,37].

### *3.2. Physico-Chemical Characteristics of the Leachates*

During the entire experimentation period, the physico-chemical characteristics were determined in the different leachates sprayed on the biopile; the results are presented in

Table 1. A high nitrogen content was noticed, mainly attributed to ammonium ions (more than 80%) and a high ionic strength with an average conductivity value of 49 mS cm$^{-1}$. The conductivity ranges of the leachates were in the range of other values reported in the literature (1.4 to 82.6 mS cm$^{-1}$) [8].

**Table 1.** Physico-chemical parameters of the leachates. Min and Max: the values measured during the trial for each parameter ($n$ = 4, except for $NO_3^-$ and $PO_4^{3-}$); and *: LOQ.

| Watering | pH | Conductivity | DOC | $NO_3^-$ | $NH_4^+$ | TDN | $PO_4^{3-}$ | VFAs |
|---|---|---|---|---|---|---|---|---|
| | - | mS cm$^{-1}$ | mg L$^{-1}$ | mgN L$^{-1}$ | mgN L$^{-1}$ | mgN L$^{-1}$ | mg$P_2O_5$ L$^{-1}$ | mg L$^{-1}$ |
| I | 8.0 | 54 | 2846 | <0.06 * | 8809 | 9670 | 4.9 | <23.4 * |
| II | 7.9 | 46 | 3323 | 134 | 11673 | 12346 | 5.2 | <23.4 * |
| III | 8.2 | 58 | 1549 | 162 | 10099 | 10686 | <0.16 * | <23.4 * |
| IV | 7.9 | 39 | 4095 | <0.06 * | 11183 | 13260 | <0.16 * | <23.4 * |
| Mean | 8 | 49 | 2953 | - | 10441 | 11490 | - | - |
| SD | 0.1 | 9 | 1068 | - | 1271 | 1615 | - | - |
| Min | 7.9 | 39 | 1549 | <0.06 * | 8809 | 9670 | <0.16* | - |
| Max | 8.2 | 58 | 4095 | 162 | 11673 | 13260 | 5.2 | - |

DOC: dissolved organic carbon; TDN: total dissolved nitrogen; and VFAs: volatile fatty acids; SD: standard deviation; LOQ: limit of quantification.

### 3.3. *Water Extract Analysis*

– Evolution of Physico-Chemical Parameters

Table 2 shows the values obtained after the physico-chemical analysis of the water extracts (WEs) throughout the experiment in the test biopile. The results show that the biopile WEs were impacted by the different leachate inputs. Although the pH remained close to neutral (7.8), we noted a strong increase in the average conductivity. The latter doubled at the end of the trial to 4 mS cm$^{-1}$, compared to 2 mS cm$^{-1}$ before the first addition of leachate. In addition, the detection of volatile fatty acids was only achieved in two samples that were collected 2 weeks after watering (Supplementary Materials, Table S2). Their presence suggests that anaerobic areas could sometimes exist in the biopile [38], but that most of the time, aerobic conditions are predominant.

**Table 2.** Physico-chemical data (±CI, $n$ = 3) of the water extracts (WEs) from the test biopile substrate. Min and Max: the values measured during the trial for each parameter ($n$ = 12, except for VFAs and before supply); and *: LOQ.

| Sampling Campaigns | pH | Conductivity | DOC | $NO_3^-$ | $NH_4^+$ | TDN | $PO_4^{3-}$ | VFAs | $SUVA_{254}$ | $E_2/E_3$ |
|---|---|---|---|---|---|---|---|---|---|---|
| | - | mS cm$^{-1}$ | mg g$^{-1}$ DW | mgN g$^{-1}$ DW | mgN g$^{-1}$ DW | mgN g$^{-1}$ DW | mg$P_2O_5$ g$^{-1}$ DW | mg g$^{-1}$ DW | L mgC$^{-1}$ m$^{-1}$ | - |
| Before first supply | 7.7 ± 0.26 | 2 ± 0.2 | 15.00 ± 12.62 | 0.027 ± 0.032 | 0.32 ± 0.16 | 1.34 ± 0.75 | 0.107 ± 0.050 | <0.4 * | 1.41 ± 0.70 | 4.71 ± 0.11 |
| After 8 weeks | 7.9 ± 0.15 | 4 ± 1.3 | 8.85 ± 9.68 | 0.080 ± 0.110 | 6.30 ± 6.07 | 9.54 ± 2.84 | 0.083 ± 0.021 | 1.5 | 2.55 ± 2.28 | 4.47 ± 0.48 |
| Mean | 7.8 | 3.6 | 7.03 | 0.061 | 6.33 | 7.94 | 0.115 | - | 2.16 | 4.53 |
| SD | 0.1 | 0.7 | 1.72 | 0.051 | 2.50 | 3.21 | 0.041 | - | 0.50 | 0.48 |
| Min | 7.6 | 3 | 4.75 | 0.003 | 3.10 | 4.02 | 0.052 | <0.4 * | 1.62 | 3.28 |
| Max | 8.0 | 5 | 9.33 | 0.159 | 10.34 | 13.18 | 0.199 | 9.2 | 3.55 | 5.17 |

DOC: dissolved organic carbon; TDN: total dissolved nitrogen; $SUVA_{254}$: specific ultraviolet absorbance at 254 nm; $E_2/E_3$: absorbance ratio at 250 nm/365 nm; and VFAs: volatile fatty acids; CI: confidence interval; SD: standard deviation; LOQ: limit of quantification.

The weak variation in the DOC values in Figure 1 shows that biopile enrichment with leachate spraying is negligible, although the leachates contained a high organic carbon content (average concentration equal to 2953 mg L$^{-1}$, Table 1). No significant differences were demonstrated between the initial and final DOC values for the test biopile, thus demonstrating a lack of accumulation of organic matter (non-parametric Mann–Whitney test,

$p > 0.05$), as well as between both biopiles (non-parametric Mann–Whitney test, $p > 0.05$). Furthermore, a decrease in the DOC values was observed after watering with water or leachate, which suggests that hydrosoluble organic compounds present in the biopiles could be washed off during the watering. Indeed, during the biodegradation phase of the organic matter from both the substrate and leachate present in the biopile, some organic compounds such as simple sugars and organic acids were produced [39] and could then be brought into the biopile core during watering. Moreover, the concentrations of nitrate and phosphate ions were not affected by the leachate additions.

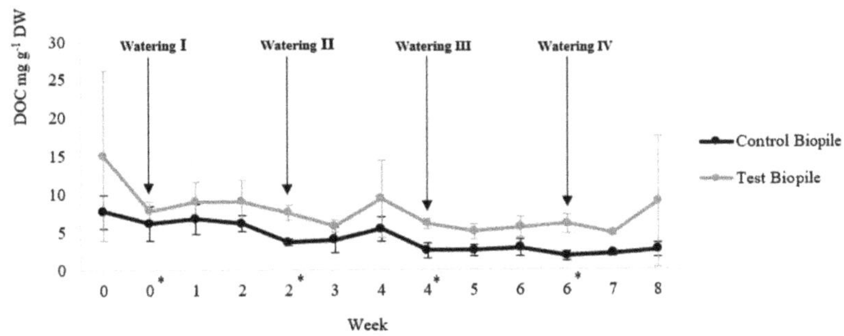


**Figure 1.** Temporal evolution of the dissolved organic carbon (DOC) concentration in the water extracts; * means after watering and the error bars represent standard deviations ($n = 3$).

From the qualitative indices determined by spectrophotometry, an evolution of the organic matter quality during the 8 weeks of testing could be observed. Indeed, the $E_2/E_3$ index varied differently for the watering steps: a non-significant decrease took place immediately after the waterings and a slight increase was observed during the following 2 weeks. Such behavior could be correlated with the previous results on DOCs, with the leaching of simple organic compounds resulting in higher $E_2/E_3$ index values during the waterings, and the following production of organic compounds by biodegradation leading to a slight increase in this ratio.

The $SUVA_{254}$ index did not vary significantly during the experiment. This index is mainly related to the degree of aromaticity and the molecular weight of the organic matter, and therefore the stabilization of humic substances [40]. It is quite normal that, at this stage of evolution, after 8 weeks of testing and with still-active biodegradation in the biopile, no trend could be observed.

These results demonstrate the very low impact of the leachate on the overall composition of the biopile. They also demonstrate that it is possible to implement a simple process on site that is capable of optimally processing leachate without considerably enriching the green waste mixture.

– Particular Evolution of Ammonium

Nitrogen compounds, and particularly ammonium ions, are of great interest due to their major contributions to the treatment of sewage sludge compost leachates. Indeed, more than 80% of the nitrogen introduced by the leachate corresponded to ammonium ions (Table 1). Figure 2a represents the temporal evolution of ammonium during the trial. We observed that the control biopile, watered with clean water, contained very little TDN. The test biopile, watered with leachate, was impacted by successive supplies of leachate with a high ammonium load. The evolution of ammonium during the experiment showed a decrease during the two weeks following a watering.

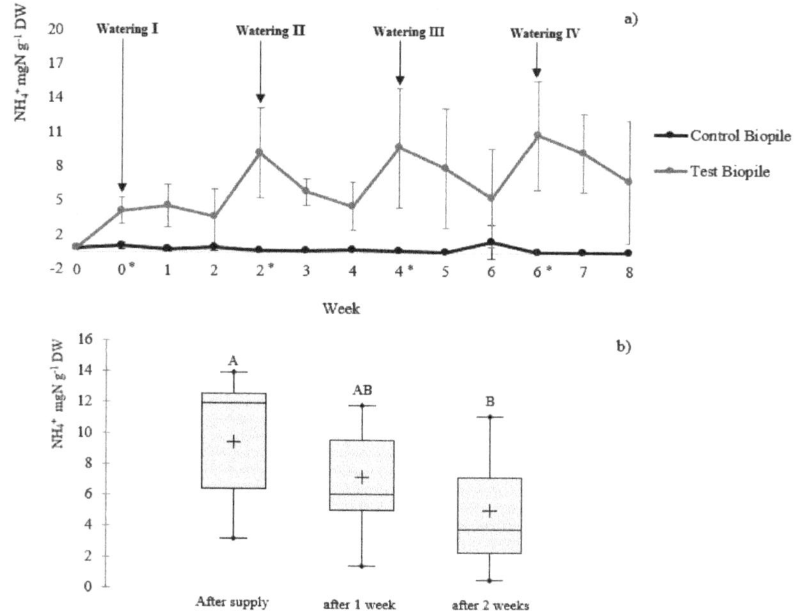


**Figure 2.** (**a**) Temporal evolution of the ammonium nitrogen ($NH_4^+$) concentration in the water extracts, and (**b**) monitoring of the ammonium nitrogen ($NH_4^+$) abatement evolution in the test biopile (* means after watering and the error bars represent standard deviations ($n$ = 3)). Different letters above boxplots indicate a significant difference between plots according to the Kruskal–Wallis test with Dunn's procedure, $p$ value = 0.05.

To underline the effects of successive leachate additions on the ammonium ion concentrations, a Kruskal–Wallis test with Dunn's procedure was undertaken; the results are presented in Figure 2b (except for the first one, as explained previously). Thus, the reduction in the ammonium nitrogen load is very significant 2 weeks after each watering. This decrease can be explained either by: (1) biological effects related to the microbial growth conditions inside the windrow (Section 3.1), or (2) the loss of ammonia nitrogen by volatilization linked to the elevated temperatures (>75 °C) reached in the biopile (Section 3.4).

### 3.4. Solid Sample Analysis

– Evolution of Physico-Chemical Parameters

Table 3 presents the physico-chemical parameters measured in the solid samples collected in the test biopile during the entire experiment. The DW decreased with each successive watering, from 69% RW to 52% RW on average during the whole campaign. No specific enrichment of the biopile with metallic trace elements was noticed during the trial, with the initial concentration values (before supply) being very close to those measured after 8 weeks. This phenomenon was also observed for the total phosphorus.

**Table 3.** Physico-chemical data (±CI, $n = 3$) of solid samples from the biopile substrate; Min and Max: the values measured during the trial for each parameter ($n = 12$, except for TP, TMEs, and before the supply).

| Sampling Campaigns | TC mg g$^{-1}$ DW | TN mgN g$^{-1}$ DW | TC/TN | $\delta^{15}$N | TP mg$P_2O_5$ g$^{-1}$ DW | TMEs (mg kg$^{-1}$ DW) | | | | | | | DW |
|---|---|---|---|---|---|---|---|---|---|---|---|---|---|
| | | | - | ‰ | | As | Cd | Cr | Cu | Ni | Pb | Zn | % RW |
| Before first supply | 324 ± 38 | 24 ± 3 | 13 ± 13 | 6.7 ± 0.9 | 8.2 ± 1.6 | 1.1 ± 0.4 | 0.3 ± 0.1 | 17.9 ± 4.7 | 44.0 ± 9.3 | 9.7 ± 2.4 | 19.1 ± 4.0 | 120.7 ± 25.1 | 69 ± 11 |
| After 8 weeks | 345 ± 12 | 24 ± 4 | 15 ± 3 | 5.2 ± 0.5 | 9.0 ± 0.6 | 1.5± 0.5 | 0.3 ± 0.1 | 16.9 ± 3.0 | 49.6 ± 5.2 | 10.0 ± 1.2 | 23.1 ± 7.3 | 133.2 ± 7.2 | 47 ± 8 |
| Mean | 328 | 25 | 13 | 6.5 | 8.4 | 1.3 | 0.3 | 20 | 46 | 9.8 | 20 | 124 | 52 |
| SD | 9.12 | 2.93 | 1.56 | 0.7 | 0.6 | 0.2 | 0.05 | 4.7 | 3.0 | 0.2 | 3.3 | 7.9 | 8.0 |
| Min | 317 | 21 | 10 | 5.2 | 8.0 | 1.1 | 0.2 | 16.9 | 44.0 | 9.7 | 16.6 | 119 | 42 |
| Max | 345 | 32 | 16 | 7.4 | 9.0 | 1.6 | 0.3 | 25.5 | 49.6 | 10.0 | 23.1 | 133 | 62 |

TC: total carbon; TN: total nitrogen; TP: total phosphorus; TMEs: trace metal elements; DW: dry weight; and RW: raw weight; CI: confidence interval; SD: standard deviation.

Concerning the total carbon, its concentration remained the same during the study. This was expected, since the organic carbon input from the leachate in a green waste windrow was negligible compared to the organic carbon already present in green waste substrates. The measured carbon was essentially organic compounds derived from fresh plant waste.

The C/N ratio was followed, as it is usually indicative of the stabilization and maturation of the composting product [41] and it gives a qualitative evaluation of the organic matter decomposition [42]. Herein, the C/N ratio remained constant with an average of 13, which was suitable for further composting conditions even after the leachate inputs [43].

– Particular Evolution of Total Nitrogen

The evolution of the total nitrogen concentration in the test biopile showed an increase immediately after the leachate inputs (Figure 3a) and a significant decrease 1 week later (Figure 3c), except after the first watering. Moreover, in Figure 3b,d, the nitrogen isotopic signature ($\delta^{15}$N) confirmed this trend by increasing after the leachate inputs (average value 7.1‰), and decreasing 2 weeks later (average value 5.5‰) (Supplementary Materials, Table S3).

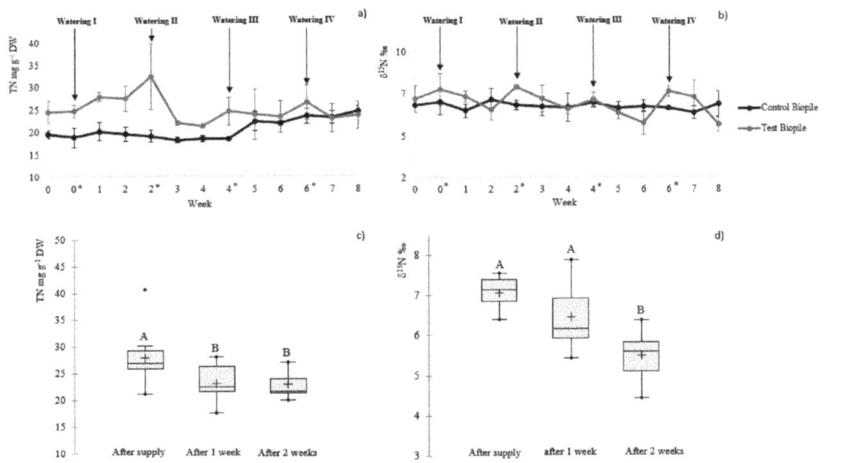


**Figure 3.** (**a**) Temporal evolution of the total nitrogen (TN) concentration; (**b**) temporal evolution of the isotopic signature ($\delta^{15}$N) (* means after watering, and the error bars represent standard deviations ($n = 3$)); (**c**) monitoring of the total nitrogen (TN) abatement evolution in the test biopile; and (**d**) the isotopic signature ($\delta^{15}$N) ($n = 3$) evolution in the test biopile. Different letters above boxplots indicate a significant difference between plots according to the Kruskal–Wallis test with Dunn's procedure, (**a**–**c**) $p$ value = 0.05, (**d**) $p$ value = 0.001.

The change in the isotope ratio of the total N clearly suggests that a loss occurred from the liquid pool of N, since a loss from solids is not affected by Rayleigh isotope enrichment in the remining pool.

The loss of nitrogen decreased with ammonium nitrogen. Our calculation with the PHREEQC model (Supplementary Materials, Figure S1) showed that, at a biopile temperature of 75 ± 5 °C, $NH_3$ accounted for about one third of the total ammonia and that this percentage reached 40% at 85 °C. The volatilization of $NH_3$ from an aqueous solution is a process that is known to cause fractionation of $\delta^{15}$ N [44]. Therefore, the loss of total nitrogen following the two weeks after watering was probably caused mainly by the volatilization of $NH_3$ [45]. It should be noted that losses of nitrogen in its gaseous form ($NH_3$) occur in the various nitrogen transformation processes [46].

– Persistent Organic Pollutant Analysis

The analysis of seven PCBs and three PAHs in the solid phase of the test biopile, monitored with the French standard (NFU 44-095), showed low levels of organic pollutants before and after 8 weeks. Regardless of the sample, no detectable PCBs were observed (all concentrations were under the detection limits). All three PAHs were detected, but only two of them were quantified, with benzo(a)pyrene being present under its quantification limit (Supplementary Materials, Table S4). The total concentration of fluoranthene and benzo(b,k)fluoranthene before and after 8 weeks of spraying were approximately the same—around 120–130 µg kg$^{-1}$. The concentration levels determined for the PAHs were equivalent to those found in the scientific literature [47]. These results demonstrate that the biopile was not contaminated by PAHs or PCBs from the leachate spraying.

### 3.5. Microbiological Analyses

The microbiological analyses showed no temporal evolution from week 0 (W0) to week 4 (W4), with a similar catabolic diversity (Figure 4a). This suggests that the leachate inputs did not significantly modify the catabolic diversity in the biopile for these first 4 weeks. From the fourth week onwards, the diversity significantly decreased compared to W0 in the control biopile. In addition, the diversity in the test biopile significantly decreased by the sixth week compared to W0. This could be related to the addition of organic matter from week 0 (W0) to week 4 (W4), which may favor the maintenance of certain microbial communities in the test biopile.

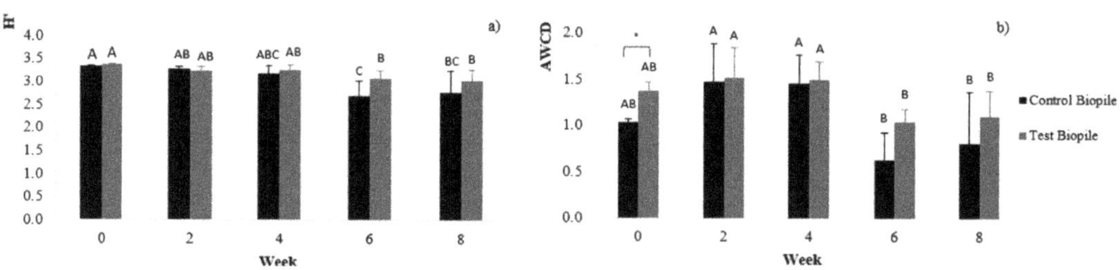


**Figure 4.** (**a**) Evolution of the Shannon–Weaver index (H′, $n$ = 3), and (**b**) evolution of the global microbial activity (AWCD: average well color development, $n$ = 3). Different letters indicate a significant difference between plots according to the Kruskal–Wallis test with Dunn's procedure; $p$ value = 0.05 for (**a**); $p$ value > 0.05 for (**b**); and * indicates a significant $t$-test result, $\alpha = 0.001$.

The overall microbial activity (Figure 4b) in week 6 decreased significantly from that measured at week 4 for both types of biopiles. However, there was no change in the global trend of microbial activities from the beginning to the end of the experiment. A significant difference at W0 between the control and test biopiles was noted before the first watering. The windrows that were put in place a few days before the start of the test could explain this result. However, after the first watering, the respective microbial activities of the two biopiles were similar. We can therefore identify two evolutionary trends during our campaign: a first phase, with high catabolic diversity and microbial activities, and a second, with lower catabolic diversity and microbial activities.

A principal component analysis [48,49] of the biological data is presented in Figure 5 according to PC1 and PC2, which supported 69% of the variance. The PCA clearly shows a difference in the catabolic profiles of the microbial communities [50] for both the control and test biopiles, depending on the time of incubation (W0 to W8) and according to PC1 (49% of variance). The projections from the four first weeks of incubation (on the positive part of the axis) were indeed discriminated from the projections of W6 and W8 (on the negative part of the axis). Moreover, it should be noted that the microbial catabolic profiles between the control and test biopiles also differed according to PC1. At W0, the microbial communities present in the test biopile were involved in the transformation of substrates such as 2-hydroxybenzoic acid, which is typical of a lignocellulosic substrate. However, after spraying, it was mainly compounds from carbohydrate families and amino acids which were the most degraded during campaigns W2 and W4. The microbial communities involved in the degradation of labile substrates in the biopile were potentially introduced and/or favored by the leachates. No marker was strongly and positively correlated to the projections from campaigns W6 and W8. This can be explained by a lower catabolic diversity and microbial activity.

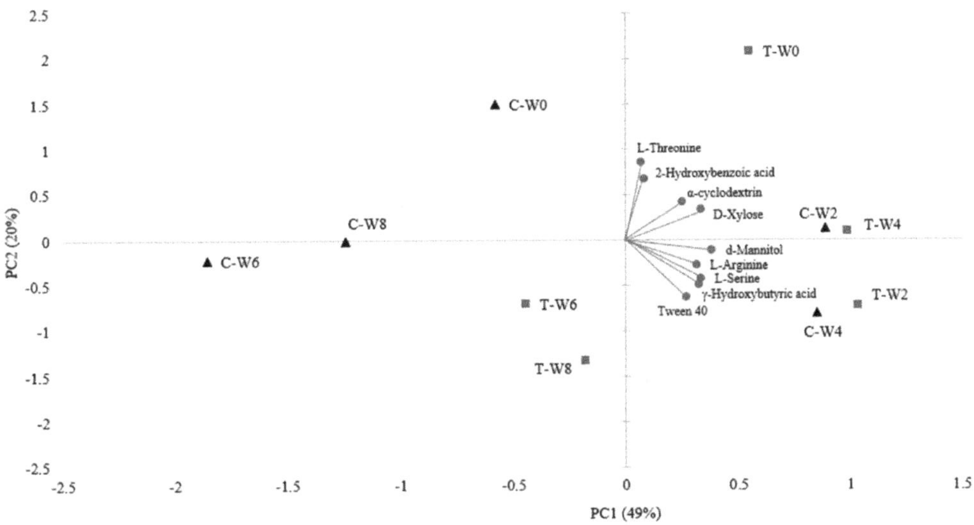


**Figure 5.** Biplot of the principal component analysis (PCA) scores, in the plane of PC1 and PC2, obtained from the control biopile (triangle) and test biopile (square) before each supply of leachate (W: Week). Substrates with stronger correlations are specified in the figure.

### 3.6. Assessment of Biopile Efficiency in the Perspective of Its Valorization

The evolution of microbial populations during the trial indicated that indigenous micro-organisms, which are responsible for the biological degradation within the biopiles (Figure 4), are well-adapted to degrade the available substrates and those potentially introduced by the leachates.

In the solid phase, among the physico-chemical parameters monitored in the French standard (NFU 44-095) which governs the production of sewage sludge composts, the TME and POP results showed that the biopile was not enriched by leachate additions (Supplementary Materials, Tables S3 and S4). In terms of agronomic criteria, the total nitrogen and total phosphorus thresholds measured in the leachate-watered biopile were similar to those measured in the control biopile (Figure 6).

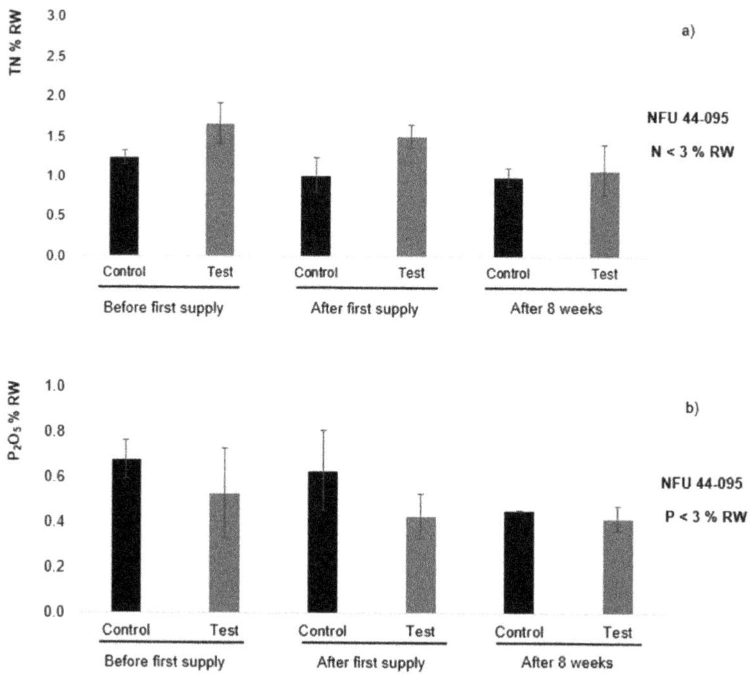


**Figure 6.** The agronomic parameters (**a**) TN (% RW) and (**b**) $P_2O_5$ (% RW) measured in the two biopiles: Control (biopile watered with clean water) and Test (biopile watered with leachate), $n = 3$.

These results validate the process of the elimination of compost leachates by spraying them on a green waste biopile. As the biopile substrate must ultimately join the composting process for its on-site valorization in a circular economy, it has been shown to meet the standards of use for MIATE composts (i.e., in accordance with the NFU 44-095 standard for monitored pollutants and agronomic parameters).

From an economical point of view, the current annual cost for the composting platforms used to treat leachates in a sewage treatment plant is around EUR 130,000. The alternative process described in this study would have a much lower cost, essentially equaling the cost of purchasing the leachate sprinkling device (EUR 20,000 for an autonomous watering device for a biopile). The material used as a substrate for the biopile is available on the composting platform, and the completion of the biopile would be carried out by staff internal to the company and would require only the completion time (about one day every 2 months for the management of a biopile).

## 4. Conclusions

This study allowed us to propose an on-site alternative approach for the treatment of leachates from sewage sludge composts on a green waste biopile with zero discharge (no percolate). The microbiological analysis indicated that the microbial populations already present in the biopile were able to ensure the biodegradation, and that the leachate input did not significantly modify the catabolic diversity of the biopile. The monitoring of the physico-chemical parameters showed that the leachate inputs did not significantly enrich the biopile with organic or inorganic elements or compounds (total organic carbon, phosphate, total phosphorus, TMEs, or POPs). A significant total nitrogen abatement in the solid phase was observed after each leachate addition every 7 days, and in the water extracts every 15 days, resulting in ammoniacal nitrogen loss.

The duration of the leachate treatment on a biopile could be improved and increased while respecting the MIATE compost product standards of use, as the biopile substrate should ultimately reach the composting process for its on-site valorization from a circular economy perspective. Another point of interest for this method of valorization is the optimization of the composting process. Future work should involve experiments with the introduction of the biopile substrate at the start of the bio-oxidation process in a certain proportion, in order to determine the potential impacts of the microbiological input on the effectiveness of the biodegradation of organic matter. We hope that the maturation phase would also be impacted and reduced over time, which would be a positive point in the management of the composting platform.

Thus, composting platforms could treat their leachates on site with a simple and rustic method that is economically viable and sustainable, and this could be expanded to other waste-composting process (household, agricultural, or industrial biowastes) that produce leachates.

**Supplementary Materials:** The following supporting information can be downloaded at: https://www.mdpi.com/article/10.3390/pr10061196/s1: Table S1: Watering schedule and sampling campaigns during the trial; the numbers represent the number of weeks, the roman numerals represent watering events, and * means after watering; Table S2: Physico-chemical data ($\pm$ CI, $n = 3$) of the water extracts (WEs) from the test biopile substrate; CI: confidence interval; SD: standard deviation; Min and Max: the values measured during the trial for each parameter ($n = 12$, except for VFAs and before supply); and *: LOQ (limit of quantification); Table S3: Physico-chemical data ($\pm$ CI, $n = 3$) of the solid samples from the biopile substrate; CI: confidence interval; SD: standard deviation; and Min and Max: the values measured during the trial for each parameter ($n = 12$, except for TP, TMEs, and before the supply); Figure S1: Speciation of dissolved $NH_4^+$ (0.5 M) as a function of temperature, calculated using PHREEQC (USGS), Version 3.7 with the database Phreeqc.dat.; Table S4: Persistent organic pollutants (Mean $\pm$ CI, $n = 5$) of solid samples from the biopile substrate; CI: confidence interval; *: LOD (limit of detection); and **: LOQ (limit of quantification).

**Author Contributions:** Conceptualization, methodology, validation, investigation, and resources, C.I.I., P.W.-W.-C., P.P. and F.T.; writing—original draft and writing—review and editing, P.H., A.-M.F.D.S., L.A., D.G., A.D., L.V., C.D., C.I.I., P.W.-W.-C., P.P. and F.T.; visualization and supervision, P.W.-W.-C., P.P. and F.T. All authors have read and agreed to the published version of the manuscript.

**Funding:** This research was funded by SARL Biotechna and the National Association for Research and Technology (ANRT), grant number 2018/0534.

**Institutional Review Board Statement:** Not applicable.

**Informed Consent Statement:** Not applicable.

**Data Availability Statement:** Data from this study can be made available upon request.

**Acknowledgments:** The authors would like to thank SARL Biotechna for having made this work possible. This work was financially supported by SARL Biotechna and the National Association for Research and Technology (ANRT). The authors extend their thanks to Anastasia Lopez for her support on the Biotechna site; to members of Laboratory LCE (Ravier Sylvain, Stéphanie Lebariller, Fabien Robert-Peillard, and Bruno Coulomb) and the IMBE Laboratory (Lisa Foli and Magalie Claeys-Bruno) for their technical help; and to Claude Nervi for his contribution of knowledge in the field of composting.

**Conflicts of Interest:** The authors declare no conflict of interest.

## References

1. EUR-Lex Official Journal of the European Union. Available online: https://eur-lex.europa.eu/legal-content/EN/TXT/HTML/?uri=CELEX:32008L0098&from=EN (accessed on 9 September 2021).
2. Clark, J.H.; Farmer, T.J.; Herrero-Davila, L.; Sherwood, J. Circular Economy Design Considerations for Research and Process Development in the Chemical Sciences. *Green Chem.* **2016**, *18*, 3914–3934. [CrossRef]

3. Ministry of Ecological Transition. Lancement de la Concertation sur le Plan National de Prévention des Déchets. Available online: https://www.ecologie.gouv.fr/lancement-concertation-sur-plan-national-prevention-des-dechets (accessed on 9 September 2021).
4. Cortés, A.; Oliveira, L.F.S.; Ferrari, V.; Taffarel, S.R.; Feijoo, G.; Moreira, M.T. Environmental Assessment of Viticulture Waste Valorisation through Composting as a Biofertilisation Strategy for Cereal and Fruit Crops. *Environ. Pollut.* **2020**, *264*, 8. [CrossRef] [PubMed]
5. Van Gestel, K.; Mergaert, J.; Swings, J.; Coosemans, J.; Ryckeboer, J. Bioremediation of Diesel Oil-Contaminated Soil by Composting with Biowaste. *Environ. Pollut.* **2003**, *125*, 361–368. [CrossRef]
6. AFNOR NF U44-095. Available online: https://www.boutique.afnor.org/fr-fr/norme/nf-u44095/amendements-organiques-composts-contenant-des-matieres-dinteret-agronomique/fa107372/520 (accessed on 9 September 2021).
7. Renou, S.; Givaudan, J.G.; Poulain, S.; Dirassouyan, F.; Moulin, P. Landfill Leachate Treatment: Review and Opportunity. *J. Hazard. Mater.* **2008**, *150*, 468–493. [CrossRef]
8. Roy, D.; Benkaraache, S.; Azaïs, A.; Drogui, P.; Tyagi, R.D. Leachate Treatment: Assessment of the Systemic Changes in the Composition and Biodegradability of Leachates Originating in an Open Co-Composting Facility in Canada. *J. Environ. Chem. Eng.* **2019**, *7*, 103056. [CrossRef]
9. Légifrance Arrêté du 22 Avril 2008 Fixant les Règles Techniques Auxquelles Doivent Satisfaire les Installations de Compostage Soumises à Autorisation en Application du Titre Ier du Livre V du Code de L'environnement—LégifranceSURAM. Available online: https://www.legifrance.gouv.fr/loda/id/JORFTEXT000018800981/ (accessed on 9 September 2021).
10. Fatima, S.; Jehangir, A.; Bhat, G.A.; Scholar, M.P. Treatment of Landfill Leachate Using Reverse Osmosis and Its Potential for Reuse. *Int. J. Innov. Res. Sci. Eng. Technol.* **2017**, *6*, 8. [CrossRef]
11. Zhao, R.; Xi, B.; Liu, Y.; Su, J.; Liu, S. Economic Potential of Leachate Evaporation by Using Landfill Gas: A System Dynamics Approach. *Resour. Conserv. Recycl.* **2017**, *124*, 74–84. [CrossRef]
12. Deng, Y.; Englehardt, J.D. Electrochemical Oxidation for Landfill Leachate Treatment. *Waste Manag.* **2007**, *27*, 380–388. [CrossRef]
13. Robinson, T.; Robinson, H. The Use of Reed Beds for Treatment of Landfill Leachates. *Detritus* **2018**, *3*, 124–140. [CrossRef]
14. Wojciechowska, E.; Obarska-Pempkowiak, H. Performance of Reed Beds Supplied with Municipal Landfill Leachate. In *Wastewater Treatment, Plant Dynamics and Management in Constructed and Natural Wetlands*; Springer: Dordrecht, The Netherlands, 2008; pp. 251–265. ISBN 978-1-4020-8235-1.
15. Banch, T.J.H.; Hanafiah, M.M.; Alkarkhi, A.F.M.; Amr, S.S.A.; Nizam, N.U.M. Evaluation of Different Treatment Processes for Landfill Leachate Using Low-Cost Agro-Industrial Materials. *Processes* **2020**, *8*, 111. [CrossRef]
16. Mouloubou, O.R.; Prudent, P.; Mounier, S.; Boudenne, J.-L.; Abaker, M.G.; Theraulaz, F. An Adapted Sequential Chemical Fractionation Coupled with UV and Fluorescence Spectroscopy for Calcareous Soil Organic Matter Study after Compost Amendment. *Microchem. J.* **2016**, *124*, 139–148. [CrossRef]
17. Chevremont, A.-C.; Boudenne, J.-L.; Coulomb, B.; Farnet, A.-M. Impact of Watering with UV-LED-Treated Wastewater on Microbial and Physico-Chemical Parameters of Soil. *Water Res.* **2013**, *47*, 1971–1982. [CrossRef]
18. Robert-Peillard, F.; Barco, E.P.; Ciulu, M.; Demelas, C.; Théraulaz, F.; Boudenne, J.-L.; Coulomb, B. High Throughput Determination of Ammonium and Primary Amine Compounds in Environmental and Food Samples. *Microchem. J.* **2017**, *133*, 216–221. [CrossRef]
19. Robert-Peillard, F.; Palacio-Barco, E.; Coulomb, B.; Boudenne, J.L. Development of a Fluorescence-Based Microplate Method for the Determination of Volatile Fatty Acids in Anaerobically Digested and Sewage Sludges. *Talanta* **2012**, *88*, 230–236. [CrossRef]
20. Park, H.-M.; Kim, Y.-M.; Lee, D.W.; Lee, S.; Lee, K.-B. Ion Chromatographic Determination of Inorganic Anions in Environmental Samples of Korea. *Anal. Sci.* **2002**, *18*, 343–346. [CrossRef]
21. Abaker, M.G.; Raynaud, M.; Théraulaz, F.; Prudent, P.; Redon, R.; Domeizel, M.; Martino, C.; Mounier, S. Rapid on Site Assessment of a Compost Chemical Stability Parameter by UV and Fluorescence Spectroscopy Coupled with Mathematical Treatment. *Waste Manag.* **2020**, *113*, 413–421. [CrossRef]
22. Hassouna, M.; Massiani, C.; Dudal, Y.; Pech, N.; Theraulaz, F. Changes in Water Extractable Organic Matter (WEOM) in a Calcareous Soil under Field Conditions with Time and Soil Depth. *Geoderma* **2010**, *155*, 75–85. [CrossRef]
23. Wang, L.; Wu, F.; Zhang, R.; Li, W.; Liao, H. Characterization of Dissolved Organic Matter Fractions from Lake Hongfeng, Southwestern China Plateau. *J. Environ. Sci.* **2009**, *21*, 581–588. [CrossRef]
24. Karanfil, T.; Schlautman, M.A.; Erdogan, I. Survey of DOC and UV Measurement Practices with Implications for SUVA Determination. *J. Am. Water Work. Assoc.* **2002**, *94*, 68–80. [CrossRef]
25. Weishaar, J.L.; Aiken, G.R.; Bergamaschi, B.A.; Fram, M.S.; Fujii, R.; Mopper, K. Evaluation of Specific Ultraviolet Absorbance as an Indicator of the Chemical Composition and Reactivity of Dissolved Organic Carbon. *Environ. Sci. Technol.* **2003**, *37*, 4702–4708. [CrossRef]
26. Gouveia, C.S.S.; Ganança, J.F.T.; Slaski, J.; Lebot, V.; Pinheiro de Carvalho, M.Â.A. Variation of Variation of carbon and isotope natural abundances (δ15N and δ13C) of whole-plant sweet potato (*Ipomoea batatas* L.) subjected to prolonged water stress. *J. Plant Physiol.* **2019**, *243*, 153052. [CrossRef]
27. Tsutaya, T.; Ogawa, N.O.; Nomura, T.; Shimizu, M.; Ohkouchi, N.; Kuze, N. Carbon and Nitrogen Stable Isotopic Offsets between Diet and Hair/Feces in Captive Orangutans. *Primates* **2021**, *62*, 945–954. [CrossRef]
28. Yousfi, S.; Serret, M.D.; Araus, J.L. Comparative response of δ13C, δ18O and δ15N in durum wheat exposed to salinity at the vegetative and reproductive stages. *Plant Cell Environ.* **2013**, *36*, 1214–1227. [CrossRef]

29. Baffi, C.; Bettinelli, M.; Beone, G.M.; Spezia, S. Comparison of Different Analytical Procedures in the Determination of Trace Elements in Lichens. *Chemosphere* **2002**, *48*, 299–306. [CrossRef]
30. Laffont-Schwob, I.; Rabier, J.; Masotti, V.; Folzer, H.; Tosini, L.; Vassalo, L.; Salducci, M.-D.; Prudent, P. Functional Trait-Based Screening of Zn-Pb Tolerant Wild Plant Species at an Abandoned Mine Site in Gard (France) for Rehabilitation of Mediterranean Metal-Contaminated Soils. *Int. J. Environ. Res. Public Health* **2020**, *17*, 5506. [CrossRef]
31. Affholder, M.-C.; Laffont-Schwob, I.; Coulomb, B.; Rabier, J.; Borla, A.; Boudenne, J.-L.; Demelas, C.; Prudent, P. Implication of Phytometabolites on Metal Tolerance of the Pseudo-Metallophyte -Rosmarinus Officinalis- in a Mediterranean Brownfield. *Chemosphere* **2020**, *249*, 126159. [CrossRef]
32. Bouhroum, R.; Boulkamh, A.; Asia, L.; Lebarillier, S.; Halle, A.T.; Syakti, A.D.; Doumenq, P.; Malleret, L.; Wong-Wah-chung, P. Concentrations and Fingerprints of PAHs and PCBs Adsorbed onto Marine Plastic Debris from the Indonesian Cilacap Coast and TheNorth Atlantic Gyre. *Reg. Stud. Mar. Sci.* **2019**, *29*, 100611. [CrossRef]
33. Kheir, M.; Roche, P.; Ziarelli, F.; Farnet da Silva, A.M. Mediterranean Coastal Conditions and Litter Type Drive Litter Microbial Responses to Drought Stress. *Eur. J. Soil Sci.* **2019**, *71*, 106–118. [CrossRef]
34. Hennink, S.; Zeven, A.C. The Interpretation of Nei and Shannon-Weaver within Population Variation Indices. *Euphytica* **1990**, *51*, 235–240. [CrossRef]
35. Mishra, A.; Nautiyal, C.S. Functional Diversity of the Microbial Community in the Rhizosphere of Chickpea Grown in Diesel Fuel-Spiked Soil Amended with Trichoderma Ressei Using Sole-Carbon-Source Utilization Profiles. *World J. Microbiol. Biotechnol.* **2009**, *25*, 1175–1180. [CrossRef]
36. Ahn, H.K.; Richard, T.L.; Glanville, T.D. Laboratory Determination of Compost Physical Parameters for Modeling of Airflow Characteristics. *Waste Manag.* **2008**, *28*, 660–670. [CrossRef] [PubMed]
37. Wéry, N.; Lhoutellier, C.; Ducray, F.; Delgenès, J.-P.; Godon, J.-J. Behaviour of Pathogenic and Indicator Bacteria during Urban Wastewater Treatment and Sludge Composting, as Revealed by Quantitative PCR. *Water Res.* **2008**, *42*, 53–62. [CrossRef] [PubMed]
38. Fang, W.; Zhang, P.; Gou, X.; Zhang, H.; Wu, Y.; Ye, J.; Zeng, G. Volatile Fatty Acid Production from Spent Mushroom Compost: Effect of Total Solid Content. *Int. Biodeterior. Biodegrad.* **2016**, *113*, 217–221. [CrossRef]
39. Thurman, E.M. Amount of Organic Carbon in Natural Waters. In *Organic Geochemistry of Natural Waters*; Developments in Biogeochemistry; Springer: Dordrecht, The Netherlands, 1985; pp. 7–65. ISBN 978-94-009-5095-5.
40. Wang; Tong, Y.; Chang, Q.; Lu, J.; Ma, T.; Zhou, F.; Li, J. Source Identification and Characteristics of Dissolved Organic Matter and Disinfection By-Product Formation Potential Using EEM-PARAFAC in the Manas River, China. *RSC Adv.* **2021**, *11*, 28476–28487. [CrossRef]
41. Guo, R.; Li, G.; Jiang, T.; Schuchardt, F.; Chen, T.; Zhao, Y.; Shen, Y. Effect of Aeration Rate, C/N Ratio and Moisture Content on the Stability and Maturity of Compost. *Bioresour. Technol.* **2012**, *112*, 171–178. [CrossRef]
42. Eiland, F.; Klamer, M.; Lind, A.-M.; Leth, M.; Bååth, E. Influence of Initial C/N Ratio on Chemical and Microbial Composition during Long Term Composting of Straw. *Microb. Ecol.* **2001**, *41*, 272–280. [CrossRef]
43. Cukjati, N.; Zupančič, G.D.; Roš, M.; Grilc, V. Composting of Anaerobic Sludge: An Economically Feasible Element of a Sustainable Sewage Sludge Management. *J. Environ. Manag.* **2012**, *106*, 48–55. [CrossRef]
44. Cejudo, E.; Schiff, S.L. Nitrogen Isotope Fractionation Factors ($\alpha$) Measured and Estimated from the Volatilisation of Ammonia from Water at PH 9.2 and PH 8.5. *Isot. Environ. Health Stud.* **2018**, *54*, 642–655. [CrossRef]
45. Sommer, S.G.; Hutchings, N.J. Ammonia Emission from Field Applied Manure and Its Reduction—Invited Paper. *Eur. J. Agron.* **2001**, *15*, 1–15. [CrossRef]
46. Peigné, J.; Girardin, P. Environmental Impacts of Farm-Scale Composting Practices. *Water Air Soil Pollut.* **2004**, *153*, 45–68. [CrossRef]
47. Lazzari, L.; Sperni, L.; Bertin, P.; Pavoni, B. Correlation between Inorganic (Heavy Metals) and Organic (PCBs and PAHs) Micropollutant Concentrations during Sewage Sludge Composting Processes. *Chemosphere* **2000**, *41*, 427–435. [CrossRef]
48. Reid, M.K.; Spencer, K.L. Use of Principal Components Analysis (PCA) on Estuarine Sediment Datasets: The Effect of Data Pre-Treatment. *Environ. Pollut.* **2009**, *157*, 2275–2281. [CrossRef]
49. Yawei, W.; Lina, L.; Jianbo, S.; Guibin, J. Chemometrics Methods for the Investigation of Methylmercury and Total Mercury Contamination in Mollusks Samples Collected from Coastal Sites along the Chinese Bohai Sea. *Environ. Pollut.* **2005**, *135*, 457–467. [CrossRef]
50. Weber, K.P.; Legge, R.L. Comparison of the Catabolic Activity and Catabolic Profiles of Rhizospheric, Gravel-Associated and Interstitial Microbial Communities in Treatment Wetlands. *Water Sci. Technol.* **2013**, *67*, 886–893. [CrossRef]

processes

MDPI

*Communication*

# A Composting Bedding System for Animals as a Contribution to the Circular Economy

**Abele Kuipers [1,*], Paul Galama [1], Lorenzo Leso [2], Kerstin Bruegemann [3] and Marija Klopčič [4]**

1 Livestock Research, Wageningen University & Research, NL-6708 WD Wageningen, The Netherlands; paul.galama@wur.nl
2 Department of Agriculture, Food, Environment and Forestry, University of Florence, IT-50145 Firenze, Italy; lorenzo.leso@unifi.it
3 Institut für Tierzüchtung, Justus-Liebig Universität Gießen, DE-35390 Gießen, Germany; kerstin.bruegemann@agrar.uni-giessen.de
4 Department of Animal Science, Biotechnical Faculty, University of Ljubljana, SI-1230 Domžale, Slovenia; marija.klopcic@bf.uni-lj.si
* Correspondence: abele.kuipers@wur.nl

**Abstract:** By-products from forestry, agriculture and nature areas are used in compost bedded-pack housing (CBP) systems for animals. In this communication, we discuss the application of a CBP system to animal farms and aspects related to the recycling and reuse of the materials in the context of a circular economy. This study is based on data from ongoing projects and literature. The following systems are discussed: (i) composting material applied to a specialized animal housing system; (ii) adding a horticultural component to the animal farm by reusing the compost, and (iii) a cooperative mixed cattle and crop farming system. The success of integrating a compost bedding component in the system depends largely on the skills of managing the composting process, the application of the material in the field, and the cost of acquiring the material. When materials are amply available, then a real contribution to the circular economy can be made. Cooperation between farmers in the utilization of by-products is another route to a more circular economy. Moreover, the analyzed systems can be seen as a Greenhouse Gases (GHG) mitigation practice because they store carbon in the soil and improve soil quality.

**Keywords:** by-products; composting material; animal housing; field application; circular economy; carbon storage

**Citation:** Kuipers, A.; Galama, P.; Leso, L.; Bruegemann, K.; Klopčič, M. A Composting Bedding System for Animals as a Contribution to the Circular Economy. *Processes* **2022**, *10*, 518. https://doi.org/10.3390/pr10030518

Academic Editor: Antoni Sánchez

Received: 26 January 2022
Accepted: 25 February 2022
Published: 5 March 2022



## 1. Introduction

There are many definitions of circular economy [1]. One definition is "an economic system of closed loops in which raw materials, components and products lose their value as little as possible, renewable energy sources are used, and system thinking is at the core". The three key Rs of sustainability are Reduce, Reuse and Recycle. Thus, the recycling and reuse of materials in agriculture contribute to a sustainable circular economy. An example is the recycling of wood remnants from sawmills, straw from grain farming, grass cuttings from nature areas and roadsides, and discarded thatch from roofs. Such materials are used in animal housing as bedding. In this communication, we focus mainly on the bedding materials used in cattle barns, especially the innovative compost bedded-pack (CBP) barns where the materials are gradually mixed with animal excreta and regularly aerated. After some time, when the bedded-pack is (semi) composted, the compost is applied to the land and reused as a soil improver.

The majority of cows in the Western world are housed in cubicle barns (free stalls; Figure 1) [2,3]. One cubicle per cow is ideal and a limited amount of bedding material is used (see Figure 1). The CBP barn (Figure 2) has a lying and movement area which, on average, is more than twice that in a cubicle barn [4]. The bedding material layer increases over time as fresh material is regularly added. The amount of bedding material used is known to be significantly higher than the quantity used in cubicle housing [4] and it gradually undergoes a composting process [5]. The CBP system for cattle was initially developed in Israel and the USA and spread, with some modification, to Europe in the last decade. Presently, it is increasing as a system in Brazil [6]. It allows cows more freedom of movement than conventional tie-stalls and cubicle barns [2–4]. The CBP is composed of a large bedded-pack, located in part of the barn and most times separated from the feed alley by a 0.5–1 m high concrete wall. A successful composting process requires the input of oxygen, so the bedded-pack is mechanically cultivated once or twice daily (see Figure 3). Additional assistance to the composting process is provided, especially in the Netherlands, by supplementary aeration of the bedded area. For this, pipes are laid on the floor under the bedded-pack (Figure 3), which facilitates either the sucking or pushing of air through the pack [3].

**Figure 1.** Cubicle barn for dairy cows.

**Figure 2.** Compost bedded-pack (CBP) barn for dairy cows.

**Figure 3.** Illustration of the stimulation of the composting process by mechanically aerating with oxygen into the pack (**left**) and the use of pipes in the concrete floor for suckling or pushing air in the bedded-pack (**right**). Source: Authors.

Most studies describe the CPB barn as animal welfare friendly (less lameness, fewer hock and leg lesions, and more natural behavior than in common housing systems) [7,8], but opinions differ on the effects on animal health. A review study [4] did not indicate real differences with respect to somatic cell count in milk from CBP compared to cubicle housing, but a study in progress using data from the FreeWalk project reported higher cell counts in the CPB system [9]. In general, a tendency towards increased longevity of the cow herd is observed while estimates of the costs of housing differ. A multi-criteria analysis in Europe estimated higher costs for the CBP system mainly because of its larger surface area per animal [10]. In contrast, US researchers reported lower costs for the CBP system [2]. Although housing construction requirements and costs differ among countries there is a general agreement that the CBP system has higher bedding costs than cubicle housing [2,10,11].

Compost bedding systems are also used in poultry husbandry [12]. Manure from poultry is drier than manure from cattle, which results in different management of the pack. Scratching of the pack by the birds serves as a natural means of aeration, in contrast with the mechanical cultivation used in CBP for cattle.

Compost production experiments in catering companies dealing with food waste [13], as well as agronomic characteristics of [14] and gaseous emissions from the CBP materials [15] were studied in view of circular processes. To our knowledge, no study has been reported linking CBP as a holistic system to the circular economy. This communication study attempts to fill that gap. Aspects of three possible CBP systems (animal based, animal plus horticulture, and mixed farming) are illustrated in the context of a circular economy. Furthermore, farmer and consumer perceptions towards the compost bedding systems are summarized, and the key practices and difficulties to the successful application of the system are synthesized and discussed on the basis of existing projects and literature sources.

## 2. Materials Used

This communication synthesizes the results from previous and ongoing studies. It is based on results from: (1) the FreeWalk project [16], with some additional analysis and observations. The FreeWalk project collected animal, climate, and nitrogen and phosphorus data from a group of study farms across six European countries. A sample of 22 CBP case farms and 22 reference cubicle farms with cattle were monitored during six visits in a two-year period (2018–2019). The case and reference farms were paired based on similar characteristics and region. Farmer and consumer perceptions of the two housing systems in the participating countries were assessed using surveys during 2018. In total, 78 farmers from six European countries, all known to are working or acquainted with the two housing systems, and 3693 consumers from eight European countries, participated in the respective surveys. (2) First outcomes of the Climate Care Cattle farming project [17], which monitors 60 innovative farms in Europe in relation to carbon management and methane and ammonia emissions from the barns and fields. (3) Information about the management of bedding materials in poultry housing. (4) Literature sources (39), which supported describing the merit of compost bedding material systems for a circular economy.

## 3. Practices and Synthesis of Results

### 3.1. Circular Systems of Applying Bedding Material

Three such systems are described:

#### 3.1.1. Application on a Dairy Farm

Wood chips and sawdust, among other materials, are used to develop the bedded-pack of, on average, 15 m$^2$ per cow with a rather wide variation between countries [2–4]. The pack is mechanically aerated by use of a tractor with tillers. Additional aeration is provided by pipes installed to stimulate the composting process [3]. In spring, the exhausted pack material is transported and superficially spread on the land (Figure 4).

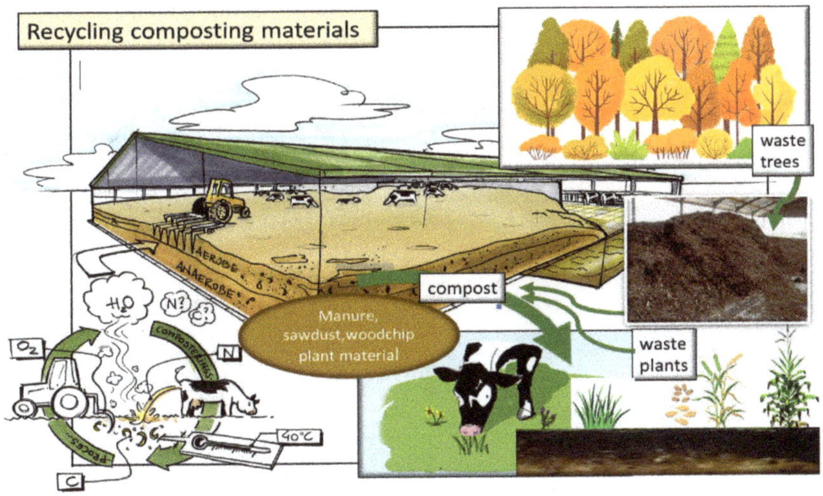


**Figure 4.** Dairy farm with compost bedded-pack and its application in the field. Source: Authors.

#### 3.1.2. Application on a Poultry Farm

In Central Europe, corn spindle granules as left over from the maize harvest, among other materials, are used as bedding. In a fattening period of 28–40 days, 1 kg bedding/m$^2$ will end up as 0.8 kg bedding/kg live weight at a stocking density of 18–20 birds/m$^2$ [12]. Scratching of the pack by the birds serves as a natural means of aeration, while moisture is supplied via the ventilation system to assist the composting process, which can start as early as 10 days after installation. Problems arise if slab formation occurs due to coccidial disease or another reason, which means the animals cannot scratc anymoreh. Animal welfare guidelines require the litter to be degradable. Mulching is a possibility to prevent or solve the formation of slabs (Figure 5). Although some farms greatly appreciate hard, flexible litter, others find that too much of it is eaten as ballast by the laying hens [12]. The resulting compost from these farms can be easily marketed.

**Figure 5.** Mulching of slabs in a broiler housing facility. Source: Hiller, P., 2018 [12].

#### 3.1.3. Added Horticultural Activity Utilizing Composted Bedded Materials

During the grazing season when the cows are at pasture, the CBP barn can be used to grow horticultural products, such as tomatoes and lettuce (Figure 6). The building has a horticultural green-house type roof that allows sunlight to enter. A layer of about 10 cm of sandy soil is spread over the bedded-pack, on which seedlings can be planted and the underlying compost supports plant growth [11]. This approach can lead to the sale of a mixture of agricultural products such as milk and vegetables from the one farm and the production system is readily visible to visiting buyers of the produce. However, the system also carries risks. Firstly, the addition of sand to the bedding and its later removal is a tedious chore that must be repeated each year; secondly, the horticultural component of the system requires additional temporary labor and skills [11].

**Figure 6.** Utilizing a compost bedded-pack barn for growing horticultural products. Source: Farm Veld & Beek, Doorwerth, the Netherlands.

#### 3.1.4. Cooperative Crop–Cattle Farm Exchange of Composting Materials and Manure

European agriculture has seen a continuous trend of intensification and specialization with the result that mixed farming systems now cover only 14% of the total agricultural area [18]. Intensification has advantages in economies of scale (efficiency of labor and equipment) but can also have drastic environmental consequences (on water and soil quality, on the atmosphere and biodiversity). Market volatility has increased over the years, and alongside climate change this threatens the resilience of specialized farms. For these reasons, "new" farming systems that adopt circular approaches are advocated with a view to further improve efficiency, while reducing their climate and environmental impacts [18].

One such a system involves closing the mineral loop by the exchange of materials and manure, and by the better use of manure (Figure 7). The mixed farming system fits this concept [19]. For instance, crop farming allows the composted materials to be incorporated into the soil. Collaboration on a local level will also play a crucial role in integrating different farming resources together with crop and livestock farmers, as illustrated in Figure 7.

**Figure 7.** Mixed farming as a system realizing a circular economy. Source: Authors.

## 3.2. Perceptions towards Bedded-Pack Barns and Compost

### 3.2.1. Farmers' Perceptions

Eighty-seven farmers from the six European countries involved in the FreeWalk project were selected on the basis of their experience with both CBP and cubicle housing systems and asked their opinions on these systems [11]. This was done via questionnaire during group meetings, guided by the local researchers who explained the questions. A Likert scale of 1 to 7 was used to compare the systems (1 = very negative to 7 = very positive).

The farmers from the various countries expressed quite similar opinions about a set of characteristics of the two housing systems (Table 1). Those from all six countries considered the CBP system more sustainable and it offered more market opportunities than the common cubicle housing system. In addition, the composted bedding material was judged a better soil improver. However, the cost of the bedding material was considered a highly significant negative aspect of the CBP system by all farmers [11]. In Europe, prices for fresh bedding materials range from around 100 euro to over 300 euro per ton [3].

**Table 1.** Appreciation of composting bedding material in CBP versus regular manure in cubicle housings (C) by farmers from these two farm housing systems in six European countries (scores from 1 = very negative to 7 = very positive) [1,2].

| All Six Countries Total | | | | |
|---|---|---|---|---|
| | **CBP** | **C** | **CBP-C** | **Mann-Whitney Test** |
| Cost aspect | | | | |
| Cost of bedding material | 3.58 | 5.53 | −1.95 | 0.000 *** |
| Sustainability aspects | | | | |
| Bedding material and slurry as soil improver | 6.15 | 3.61 | 2.54 | 0.000 *** |
| Bedding material and slurry as fertilizer | 5.82 | 4.06 | 1.76 | 0.000 *** |
| Bacteria in bed and milk | 5.15 | 4.91 | 0.23 | |
| Smell | 5.98 | 4.53 | 1.45 | 0.000 *** |
| Marketing aspects | | | | |
| Better animal life certificate | 6.05 | 3.52 | 2.54 | 0.000 *** |
| High quality dairy products | 5.69 | 4.06 | 1.63 | 0.000 *** |
| High quality manure products | 5.67 | 3.32 | 2.34 | 0.000 *** |

[1] Mann–Whitney two-tailed exact significance test was applied; *** $p < 0.001$. [2] Number of farmers participating: from the Netherlands CBP 24 and C 20, Germany CBP 7 and C 7, and other 4 countries (Austria, Italy, Slovenia, and Sweden) CBP 10 and C 10. Source: Adapted from Klopčič et al., 2021 [11].

### 3.2.2. Consumers' Perceptions

An extensive consumer survey [20] was carried out in eight European countries to compare the three most widely used cattle housing systems. The survey of random participants, which was carried out by a marketing bureau, used photographs of the various housing systems to illustrate them to the respondees. A Likert scale of 1 to 10 was used (1 = not worried at all/fully safe to 10 = very worried/not safe at all). Overall, the consumers were concerned over the safety of their food but felt that products grown using the compost from the CBP barns was safer (Table 2). Only small differences existed between countries [20]; otherwise, all of the differences mentioned here were statistically significant ($p < 0.05$). Males were less concerned than females over food safety but still felt that food from the CBP system was safer. Younger consumers were more concerned about food safety than older consumers, and across all age groups, the youngest age group felt that food grown in the barn compost was less safe. Those in urban areas were more concerned over food safety than those in suburban areas and, compared with those in rural areas, felt that the food from the compost system was less safe [20]. Some responses in focus groups indicated concerns over antibiotic seepage, bacteria in the pack, and the hygiene of the pack [11].

## 3.3. *Choices and Dilemmas*

### 3.3.1. Management of the Bedded-Pack

Two key variables in managing the pack and creating a proper composting process are pack temperature and moisture content [2]. A high temperature (45–65 °C) is considered necessary to create effective composting and material sanification [5,13,21]. The pack temperature (10 to 50 °C) and moisture content (55 to 70%) at 20 cm deep of the 22 CBP included in project FreeWalk are illustrated in Figure 8 over 12 months [22]. A higher temperature was associated with lower moisture content. Clearly, the optimal composting temperatures are not reached in practice. This is partly because the system is still in development, but also because the need is not felt, such as in the summertime when the packs are mostly good looking and dry enough. In fact, under these conditions a semi-composting process takes place.

An over-wet pack is detrimental to the composting process, and it also increases the dirtiness of the cows. Thus, managing the pack requires a particular skill set [4,5,22]. In Europe, bedded-packs tend to be drier and warmer during the summer period. However, higher temperatures most likely result in increased ammonia and methane emissions from the pack [15] and increased thermophilic bacteria activity in the pack [23].

**Table 2.** Mean scores and standard deviations for worry about (a) the safety of food consumed in general (1 = "not worried at all" to 10 = "very worried") and (b) safety of consuming food products using the compost from the compost bedded-pack housing system (1 = "fully safe" to 10 = "not safe at all") by country and in total [1,2].

| | (a) Food Safety Worry in General | | (b) Compost Food Products Safety | |
|---|---|---|---|---|
| **Variable** | **Mean** | **Std. Dev** | **Mean** | **Std. Dev.** |
| Country | | | | |
| Austria | 6.01 [cd] | 2.20 | 3.02 [bc] | 1.85 |
| Germany | 6.15 [c] | 2.18 | 3.13 [c] | 1.86 |
| Italy | 7.62 [a] | 1.79 | 3.15 [c] | 1.90 |
| Netherlands | 5.65 [d] | 2.15 | 3.21 [c] | 1.66 |
| Norway | 5.65 [d] | 2.67 | 2.60 [a] | 1.97 |
| Slovakia | 6.94 [b] | 2.22 | 3.07 [bc] | 2.00 |
| Slovenia | 7.88 [a] | 2.10 | 2.37 [a] | 2.24 |
| Sweden | 5.77 [cd] | 2.43 | 2.77 [ab] | 2.09 |
| **Total** | **6.49** | **2.36** | **2.94** | **1.96** |

[1] The different letter within each mean column section indicate the significant differences between variable options as evaluated by Tukey's HSD ($p < 0.05$). [2] The number of consumers who participated in the survey from Austria was 415; Germany 633; Italy 592; the Netherlands 423; Norway 401; Slovakia 410; Slovenia 397; and Sweden 422. Source: Adapted from Waldrop et al., 2021 [20].

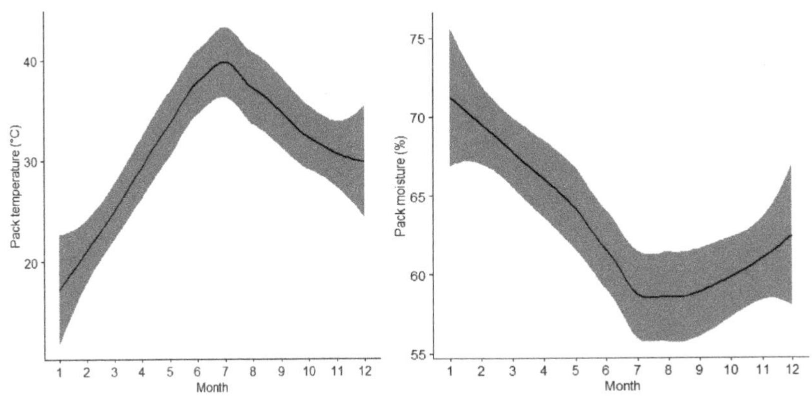


**Figure 8.** Development of the average and range in temperature (**left**) and the moisture content (**right**) at 20 cm deep in 22 composting bedded-pack's over the year. Source: Taken from Leso, L., 2021 [22].

3.3.2. Bacterial Flora

Household litter composted in bedded-pack barns at higher temperatures was found to have high levels of thermophilic aerobic sporeformers (TAS) [24]. This may threaten the quality of sterilized milk products and, consequently, its use was prohibited in barns in the Netherlands. Certain strains of mastitis bacteria also like high temperatures [25], while other strains do not [7].

The moisture content of the bedding material, the relative humidity at 0.10 m and 1.30 m above the surface, and the season of sampling, all significantly affected the level of TAS [23]. The relationship between moisture content and TAS is illustrated in Figure 9. A higher moisture content of the pack seems to be associated with less thermophilic activity, probably because of the lower temperatures and composting activity inside such a pack.

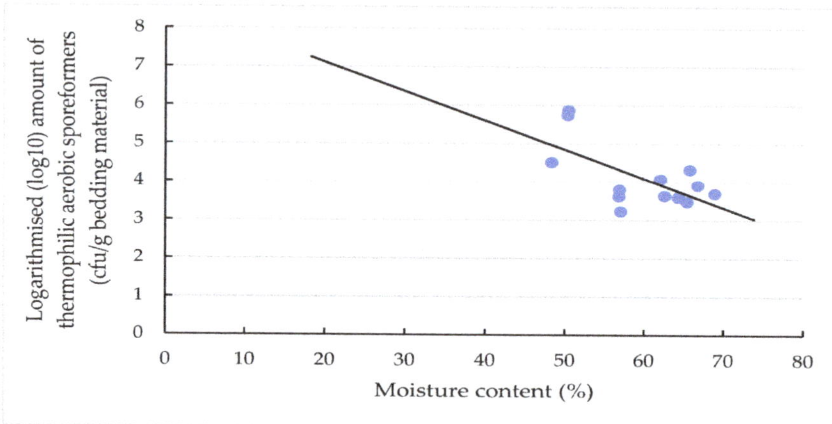


**Figure 9.** Relationship between thermophilic bacteria and moisture content in composting bedding material; $y = -0.08x + 8.62$ ($R^2 = 0.34$). Source: Taken from Giambra et al., 2021 [23].

### 3.3.3. Application in the Field

An important benefit of composted bedding material is as a source of organic matter and nitrogen fertilizer for the soil [14,26] as illustrated in Figure 10. Although compost is known to improve the soil structure and bioactivity, the composting process is more complicated than generally appreciated. When the composting of bedding materials was experimentally simulated in trials of three months duration, the stability of the organic matter did not indicate real evidence of composting [14]. Further, some farmers have claimed an increased ash content (nonorganic matter) of grass samples harvested from compost treated swards [27], but this was not observed in the survey of the farmers described above (Table 1).

**Figure 10.** Bedding material spread on grassland and arable land. Source: Authors.

The slow release of nitrogen from the bedding materials has been reported by several authors [28,29] so grassland management and fertilizer appplication may need to be adapted for the effects of compost bedding material as a fertilizer and soil structure improver.

3.3.4. Opportunities

The product from the successful composting of bedding materials mixed with excreta of animals is commercially valuable. Selling the compost in bags or loose as pellets provides the opportunity for additional income [29,30]. An example is the providing of organic bedding material by a CBP farmer in Slovenia to farmers engaged in the production of vegetable and flower seedlings.

Some farmers use the heat of the composting bedded-pack to warm up their farm office or visitors' room, or to prevent the animal drinking bowls from freezing. In another case, a German farmer with both a CPB and cubicle barn reused about 75% of the CBP compost as bedding material in the cubicle stalls with the remainder used as fertilizer. It was estimated that this reduced the total bedding material costs by 80%. Afterwards, the bedding material from the cubicles was used in the farm biogas plant to produce energy. These are examples of circularity within or between farm businesses.

The relatively high cost of the bedding materials is a negative factor but new materials such as Posidonia oceanica and miscanthus [31], and grass harvested in nature areas and on roadsides [32] should be less expensive.

Within the European Green Deal Package [33] there are opportunities to increase carbon storage in soil to combat climate change. In the CBP system, carbon enters the farm via industry or agricultural by-products and is mixed with animal manure. This is then applied to the soil thereby sequestering carbon, which might otherwise be lost to the atmosphere. Carbon certificates are being considered as a tool for advancing the Green Deal objectives. An innovative farmer in the Netherlands has already deployed a novel climate mitigation practice on his farm. The air from manure storage, rich in ammonia and methane, is pushed through tubes to a special drainage system 60 cm underground from where it diffuses into the soil. The ammonia and methane are thought to be fixed by soil bacteria, although this is not yet proven.

## 4. Discussion

The EU New Deal [33], and the Horizon Europe Research Program 2021–2027 [34], together with similar worldwide programs have, as top priorities, the achievement of carbon neutrality. The recycling and reuse of organic materials, including the use of organic bedding materials in CBP systems, can help to increase carbon storage in the soil. This contributes to mitigating global warming, improves soil structure and bioactivity, and enhances biodiversity. It is enhancing a circular economy.

While in Israel and Brazil there is widespread use of CBP systems, the number of such systems in all other countries remains limited [2,3]. Much more common are the conventional 'deep straw' barns (yards) [2]. In these barns, the lying area consists of straw, which is stapled over time, but composting is not a systematic part of the system and the surface per cow is less than in CBP. Deep straw barns are most widely used for fattening cattle and suckler cows. The relatively dry manure of these animals helps to maintain a rather dry pack. The pack tends to grow in volume during the season, while in CBP, composting degrades the material to smaller particles, which makes the pack more compact [21].

Although the CBP housing system is still in development [2,3], both farmer and consumer surveys revealed positive impressions of its sustainability compared to cubicle or tie-stall barns [11,20]. The CBP system was also preferred on food safety criteria [20]. Alongside these positive impressions, however, are some concerns on the possible risk of bacteria and antibiotic transfer from the compost bedding material to the growing vegetables and horticultural products from compost treated soil [11].

The high material costs of the wood by-products and straw due to competition from bio-gas and other industries adversely affects the economic viability of the system. Accordingly, exploration of cheaper by-products and waste products will be of increasing interest [11,14,31].

The operation of a successful bedded-pack system requires particular management skills. The composting process is affected by the temperature and moisture content of the bedding, as well as the ambient temperature and humidity [4]. In temperate climates, maintaining adequate conditions and a proper composting process in the bedded-pack to achieve a stable humus-like end-product, especially during the winter, can be a severe challenge [5,14,22].

While the composting process is enhanced by higher temperatures, such temperatures also facilitate the growth of thermophilic bacteria [23,24]. Bacteria such as (X)TAS are disliked by dairy processing companies in some countries because of their possible effects on sterilized dairy products [24]. To date, research with CBP systems did not find any detectable (X)TAS in cows' milk even though detectable levels were found in the bedded-pack [35]. In addition, several strains of mastitis-causing bacteria thrive in similar conditions to that of composting bacteria and microbes [7,24,25]. This suggests that the microbiology of the bedded-pack aerobic and anaerobic processes and the possible transfer of bacteria to the animal and animal products may be complex and requires more fundamental research. Maintaining optimal animal hygienic procedures is likely to be important [25].

Combining all of the new insights gained concerning the management of the bedded-pack leads to the conclusion that the moisture and dry matter content may be the best indicator for the quality of the pack [22]. A dry pack (above 40% dry-matter) has, indeed, several advantages such as a better composting process and cleaner cows [25], affecting the quality of the end-products. However, the risk of TAS [23,24] and particular mastitis bacteria presence [25] increase with the combination of higher dry matter content and higher temperatures of the pack. The composting process and risk of TAS presence is expected to slow down when the pack reaches above approximately 65% dry matter, but more knowledge is needed to typify the ongoing processes in this dry matter range.

In addition, new insights into ammonia and carbon dioxide emissions from composting bedded-packs would be particularly beneficial in evaluating these systems. One study indicated that the critical pack Carbon:Nitrogen (C:N) ratio at which volatile N loss from the barn was zero, was 35 in line with critical ratios found in other studies [36]. The authors concluded that controlled thermophilic composting of a woodchip bedded-pack, at a relatively high C:N ratio, has the potential to minimize volatile N loss from the CBP barns. Another study [37] showed ammonia emission to be reduced by 32% but methane emission was increased by 34% compared to a reference cubicle housing system with slurry.

Organic bedding materials enriched with nitrogen and phosphorus from manure and urine is a good soil improver [26] comparable with old fashioned manure from tie-stall and deep straw barns. The CBP system aims to degrade the materials in the compost to a product ideal for spreading on grassland swards or soil. However, when the composting process is incomplete, its application to grassland may result in some large particles which take longer to degrade. This is not a problem in soil application where the compost is incorporated into the soil at some depth, as in arable farming.

A mixed animal-crop farming system can make a positive contribution to the circular economy [18,26]. Such a system can be operated either within a single farm business or by cooperation between specialised animal and arable or open field vegetable farms [19]. Its use in individual farms, or via cooperation between farms, may also reduce the import of animal protein feed from outside of the farm, thereby conserving resources and reducing fossil fuel use while contributing to replenished soil carbon reserves [19,34].

Most studies cited in this communication are on-farm based studies. Aditionally, models scaling-up from farm to regional or international level [38], including Life Cycle Analysis (LCA) studies [13,39], are recommanded to picture the whole cycle of by-products and materials. Such studies should preferably also compare composting to other processes, such as using the materials for energy production and include the complete (international) transport and marketing chain [39].

## 5. Conclusions

The recycling and reuse of by-products in agriculture can contribute to a sustainable circular economy. In this communication, we focused on the recycling of bedding materials from animal barns, especially the innovative CBP barns, and on cooperative crop-animal farm exchange of composting materials and manure. A holistic and synthesizing approach was used to evaluate the contribution of these systems to the circular economy.

Both farmer and consumer surveys revealed positive impressions of CBP's sustainability compared to cubicle or tie-stall barns. Reuse of the compost from the CBP system is also promising for horticultural production. However, managing the bedded-pack requires a particular skill set, and the high costs of wood by-products and straw can adversely affect the economic viability of the system. Accordingly, an exploration of cheaper and suitable by-products and waste products is of crucial interest.

Mixed animal-crop farming systems that foster the exchange and reuse of materials would make a positive contribution to the circular economy. The recycling and reuse of organic materials helps to increase carbon storage in soil. This contributes to mitigating global warming and improved soil structure and bioactivity. Greater insight into ammonia and carbon dioxide emissions would be beneficial in evaluating these systems. Additionally, the scaling-up of models from farm to regional, national, or international level is recommended to obtain an overview of the complete cycle of by-products and materials.

**Author Contributions:** Conceptualization, A.K., P.G., M.K.; methodology, A.K.; software, A.K.; validation, A.K.; formal analysis, A.K., P.G.; investigation, A.K., P.G., L.L., K.B., M.K.; resources, M.K., A.K.; data curation, A.K.; writing—original draft preparation, A.K.; writing—review and editing, A.K., P.G., L.L., K.B., M.K.; visualization, A.K., P.G., M.K.; supervision, A.K.; project administration, M.K.; funding acquisition, P.G., M.K. All authors have read and agreed to the published version of the manuscript.

**Funding:** This research was funded by the European Union's Horizon 2020 Research and Innovation Program under Grant Agreements No 696231 and No 696356 for research carried out within the ERA-NET SusAn projects "FreeWalk" and "CCCfarming".

**Data Availability Statement:** The own data presented in this study are available on request from the authors.

**Acknowledgments:** We wish to thank our FreeWalk partners Ulf Emanuelson and Isabel Blanco Penedo (Swedish University of Agricultural Sciences), Sven Koenig (Justus Liebig University Giessen, Germany), Jantine van Middelkoop (Wageningen Livestock Research, the Netherlands), Matteo Barbari (University of Firenze, Italy), Elfriede Offner-Schroeck and Andreas Zentner (Raumberg-Gumpenstein, Austria) and Megan Waldrop and Jutta Roosen (Technical University Munchen) for assisting in translating, distributing and/or collecting data in the partner countries, as well as the partners of the CCCfarming project for discussions contributing to the content of this communication. We are grateful to Gerry Keane, Leixlip Co. Kildare, for the editing work on this article.

**Conflicts of Interest:** The authors declare no conflict of interest.

## References

1. Kirchner, J.; Reike, D.; Hekkert, M. Conceptualizing the circular economy: An analysis of 114 definitions. *J. Resour. Conserv. Recycl.* **2017**, *127*, 221–232. [CrossRef]
2. Bewley, J.M.; Robertson, L.M.; Eckelkamp, E.A. A 100-Year Review: Lactating dairy cattle housing management. *J. Dairy Sci.* **2017**, *100*, 10418–10431. [CrossRef] [PubMed]
3. Galama, P.; Ouweltjes, W.; Endres, M.; Sprecher, J.R.; Leso, L.; Kuipers, A.; Klopčič, M. Future of housing for dairy cattle: Symposium review. *J. Dairy Sci.* **2020**, *103*, 5759–5772. [CrossRef] [PubMed]
4. Leso, L.; Barbari, M.; Lopes, M.A.; Damasceno, F.A.; Galama, P.; Taraba, J.L.; Kuipers, A. Invited review: Compost bedded-pack barns for dairy cows. *J. Dairy Sci.* **2020**, *103*, 1072–1099. [CrossRef]
5. Hernandez, A.J.; Halbach, T.R.; Widmer, S.; Weithe, A. *Media Evaluation for Composted Bedded Pack Dairy Barns*; Project, Report no. 2005-078; University of Minnesota: Minneapolis, MN, USA, 2006. [CrossRef]
6. Damesceno, F.A.; University of Lavras, Brazil. Personal communication, 2020.
7. Eckelkamp, E.A.; Taraba, J.L.; Akers, K.A.; Harmon, R.J.; Bewley, J.M. Understanding compost bedded pack barns: Interactions among environmental factors, bedding characteristics, and udder health. *Livest. Sci.* **2016**, *190*, 35–42. [CrossRef]

8. Blanco-Penedo, I.; Ouweltjes, W.; Ofner-Schröck, E.; Brügemann, K.; Emanuelson, U. Symposium review: Animal welfare in free-walk systems in Europe. *J. Dairy Sci.* **2020**, *103*, 5773–5782. [CrossRef]
9. Emanuelson, U.; Brügemann, K.; Klopčič, M.; Leso, L.; Ouweltjes, W.; Zentner, A.; Blanco-Penedo, I. Animal health in compost-bedded pack and cubicle dairy barns in six European countries. *Animals* **2022**, *12*, 396. [CrossRef]
10. Hovstad, K.A.; NIBIO, Norway. Multi-criteria-analysis. Personal communication, 2021.
11. Klopčič, M.; Erjavec, K.; Waldrop, M.; Roosen, J.; Engel, P.; Galama, P.; Kuipers, A. Consumers' and farmers' perceptions in Europe regarding the use of composted bedding material from cattle. *Sustainability* **2021**, *13*, 5128. [CrossRef]
12. Hiller, P. Einstreu in der Geflügelhaltung-Tipps und Anregungen für ein gutes Einstreumanagement. Landwirtschaftszimmer Niedersachsen. EIP-Agri-Projekt CarboFeet-Beimischung von Pflanzenkohle. 2018. Available online: https://www.lwk-niedersachsen.de/lwk/news/29729_Einstreu_in_der_Gefl%C3%BCgelhaltung-_Tipps_und_Anregungen_f%C3%BCr_ein_gutes_Einstreumanagement (accessed on 14 January 2022).
13. Stunženas, E.; Kliopova, I.; Kliaugaite, D.; Pranas Budrys, R. Industrial symbiosis for optimal bio-waste management and production of a higher value-added product. *Processes* **2021**, *9*, 2228. [CrossRef]
14. Llonch, L.; Gordo, C.; Lopez, M.; Castillejos, L.; Ferret, A.; Balanya, T. Agronomic characteristics of the compost-bedded pack made with forest biomass or sawdust. *J. Processes* **2021**, *9*, 546. [CrossRef]
15. Sayara, T.; Sanchez, A. Gaseous emissions from the composting process: Controlling parameters and strategies of mitigation. *Processes* **2021**, *9*, 1844. [CrossRef]
16. EU ERA-NET SUSAN Project FreeWalk. Available online: www.freewalk.eu (accessed on 14 January 2022).
17. EU Joint Call ERA-NET SUSAN/ERA-GAS/ICT-AGRI Project Climate Care Cattle Farming. Available online: www.CCCfarming.eu (accessed on 14 January 2022).
18. EIP-Agri Focus Group Report on Mixed Farming Systems. 2017. Available online: https://ec.europa.eu/eip/agriculture/sites/default/files/fg16_mixed_farming_final-report_2017_en.pdf (accessed on 14 January 2022).
19. Animal Task Force. Research and Innovation towards a More Sustainable and Circular European Agriculture, Exploring Synergies between the Livestock and Crop Sectors. Plants for the Future Position Paper. 2019. Available online: http://animaltaskforce.eu/Portals/0/ATF/Downloads/ATF_PlantETP_joint-paper_Sept2019.pdf (accessed on 14 January 2022).
20. Waldrop, M.E.; Roosen, J.R. Consumer acceptance and willingness to pay for cow housing systems in eight European countries. *Q Open* **2021**, *1*, qoab001. [CrossRef]
21. de Bertoldi, M.; Vallini, G.; Pera, A. The biology of composting: A review. *Waste Manag. Res.* **1983**, *1*, 157–176. Available online: https://www.fazverde.com.br/wp-content/uploads/2020/12/The-biology-of-composting.pdf (accessed on 14 January 2022). [CrossRef]
22. Leso, L.; Galama, P.; Blanco-Penedo, I.; Bruegemann, K.; Zentner, A.; Klopčič, M.; Hovstad, K.A.; Barbari, M. Bedding management in compost-bedded pack and cubicle barns for dairy cows in Europe. In *V: Book of Abstracts of the 72nd Annual Meeting of the European Federation of Animal Science: Davos, Switzerland, September 2021*; Wageningen Academic Publishers: Wageningen, The Netherlands, 2021; p. 407.
23. Giambra, I.J.; Jahan, Y.; Yin, T.; Engel, P.; Weimann, C.; Bruegemann, K.; Koenig, S. Identification of thermophylic aerobic sporeformers in bedding material of compost-bedded dairy cows using microbial and molecular methods. *Animals* **2021**, *11*, 2890. [CrossRef]
24. Driehuis, F.; Lucas-van den Bos, E.; Wells-Bennik, M.H.J. *Risico's van microbiële contaminanten van strooisels: Compost, geschieden mest, paardenmest en vrijloopstallen*; NIZO: Ede, The Netherlands, 2012.
25. Black, R.A.; Taraba, J.L.; Day, G.B.; Damasceno, F.A.; Newman, M.C.; Akers, K.A.; Wood, C.L.; McQuerry, K.J.; Bewley, J.M. The relationship between compost bedded pack performance, management, and bacterial counts. *J. Dairy Sci.* **2014**, *97*, 2669–2679. [CrossRef]
26. Middelkoop, J.C.; Van De Boer, C.H.; Galama, P.; Brügemann, K.; Leso, L.; Blanco-Penedo, I.; Zentner, A.; Klopčič, M. Characteristics of organic manure from 'Freewalk' housing, compared with slurry, and their appreciation by farmers. In *Proceedings of the 28th General Meeting of the European Grassland Federation, Helsinki, Finland, October 2020*; Grassland Science in Europe; Natural Resources Institute (Luke): Helsinki, Finland, 2020; Volume 25, pp. 686–688.
27. Bakhuizen, G.; farmer from Abcoude, The Netherlands; Wieringa, J.; Farmer from Doorwerth, The Netherlands. Personal communication, 2019.
28. Alemi, H.; Kianmehr, M.H.; Borghaee, A.M. Effect of pellet processing of fertilization on slow-release nitrogen in soil. *Asian J. Plant Sci.* **2010**, *9*, 74–80. [CrossRef]
29. Pampuro, N.; Caffaro, F.; Cavalo, E. Reuse of animal manure: A case study on stakeholders' perceptions about pelleted compost in Northwestern Italy. *Sustainability* **2018**, *10*, 2028. [CrossRef]
30. Eggerth, L.L.; Diaz, L.F.; Chang, M.T.F.; Iseppi, L. Chapter 12 Marketing of composts. *Waste Manag.* **2007**, *8*, 325–355.
31. Ferraz, P.F.P.; Ferraz, G.A.; Leso, L.; Klopcic, M.; Rossi, G.; Barbari, M. Evaluation of the physical properties of bedding materials for dairy cattle using fuzzy clustering analysis. *Animals* **2020**, *10*, 351. [CrossRef]
32. Heijbroek., W.; Project Leader Waterschap Stichtse Rijnlanden, The Netherlands. Personal communication, 2021.
33. EU Green Deal Package. 2019. Available online: https://www.switchtogreen.eu/the-eu-green-deal-promoting-a-green-notable-circular-economy/ (accessed on 14 January 2022).

34. Horizon Europe. 2021. Available online: https://ec.europa.eu/info/research-and-innovation/funding/funding-opportunities/funding-programmes-and-open-calls/horizon-europe_en (accessed on 14 January 2022).
35. Bruegemann, K.; Justus Liebig University, Giessen, Germany; Galama, P.; Wageningen University & Research, Wageningen, The Netherlands. Personal communication, 2021.
36. De Boer, H.C.; Wiersma, M. Thermophilic composting of the pack can reduce nitrogen loss from compost-bedded dairy barns. *Biosyst. Eng.* **2021**, *210*, 20–32. [CrossRef]
37. van Dooren, H.J.C.; Hol, J.M.G.; Blanken, K.; Galama, P.J. *Gasvormige Emissies Uit Vrijloopstallen Met Houtsnipperbodems*; Report 1163; Wageningen University & Research: Wageningen, The Netherlands, 2019.
38. Yang, Z.; Muhayodin, F.; Larsen, O.C.; Miao, H.; Xue, B.; Rotter, V.S. A review of composting process models of organic solid waste with a focus on the fates of C, N, P and K. *Processes* **2021**, *9*, 473. [CrossRef]
39. De Boer, I.J.M.; Van Ittersum, M.K. *Circularity in Agricultural Production*; Wageningen University & Research: Wageningen, The Netherlands, 2018; Available online: https://www.wur.nl/upload_mm/7/5/5/14119893-7258-45e6-b4d0-e514a8b6316a_Circularity-in-agricultural-production-20122018.pdf (accessed on 14 January 2022).

processes
MDPI

*Article*

# Industrial Symbiosis for Optimal Bio-Waste Management and Production of a Higher Value-Added Product

**Edgaras Stunžėnas *, Irina Kliopova, Daina Kliaugaitė and Rimas Pranas Budrys**

Institute of Environmental Engineering, Kaunas University of Technology, Gedimino St. 50, LT-44239 Kaunas, Lithuania; irina.kliopova@ktu.lt (I.K.); daina.kliaugaite@ktu.lt (D.K.); rimas.budrys@ktu.edu (R.P.B.)
* Correspondence: edgaras.stunzenas@ktu.edu; Tel.: +370-626-88384

**Abstract:** A considerable amount of food waste ends up in centralized treatment plants due to the lack of preventive measures, resulting in significant environmental impacts. Hospitality food waste management is even more resource-intensive because of animal by-products regulation. According to this regulation, companies must store and then consign waste to specific waste managers. The extensive need for transportation of high-moisture-content materials is the leading cause of the impact. Moreover, the management of category III animal by-products is costly for companies. A previous study has shown the economic benefits of decentralized animal by-product treatment by intensive composting in catering companies. Although the produced compost was characterized by exceptional quality parameters, it was phytotoxic. The investigation of hospitality waste management is scarcely discussed among scholars, and waste management on a regional scale is nearly absent. This study examines the regional management of hospitality food waste by exploiting the municipal waste management infrastructure and intensive composting at the source. The co-maturation experiment with animal by-products and municipal green waste primary composts showed that the phytotoxicity parameters of the cured compost were in the optimal range or below the thresholds (conductivity (1.1 mS $cm^{-1}$), dissolved organic carbon (82 mg $kg^{-1}$), and $NH_4^+/NO_3^-$ ratio (0.0027)). Additionally, the amounts of total nitrogen, water-soluble nitrogen, and water-soluble phosphorus in the compost were rated as very high. Finally, inventory and environmental impact analysis of the current and planned management approaches showed a reduction in 12 of 18 impact categories.

**Keywords:** animal by-products; intensive composting; industrial symbiosis; catering; compost quality parameters; hospitality; life cycle assessment; life cycle inventory

**Citation:** Stunžėnas, E.; Kliopova, I.; Kliaugaitė, D.; Budrys, R.P. Industrial Symbiosis for Optimal Bio-Waste Management and Production of a Higher Value-Added Product. *Processes* **2021**, *9*, 2228. https://doi.org/10.3390/pr9122228

Academic Editor: Antoni Sánchez

Received: 22 November 2021
Accepted: 8 December 2021
Published: 10 December 2021

**Publisher's Note:** MDPI stays neutral with regard to jurisdictional claims in published maps and institutional affiliations.



## 1. Introduction

Food waste (FW) is a major concern, and its quantity will most likely increase in parallel with population growth. The Food and Agricultural Organization of the United Nations (FAO) has forecasted that by 2050, food production will increase by over 70 % to feed 9.1 billion people [1]. This growth would definitely result in a substantial generation of FW, considering that nearly one-third of food is currently wasted [2]. Additionally, as the Boston Consulting Group states, this wastage could increase from 1.6 to 2.1 billion tons by 2030 if the current tendency does not change [3]. Although FW prevention is most preferable, only avoidable FW streams can be addressed, which represents 64% of the total FW [4]. Therefore, there is a need for conventional but sustainable end-of-pipe treatment methods.

In this regard, the hospitality sector has the potential for more sustainable management of unavoidable FW. For instance, in the European Union (EU), hotels, restaurants, cafeterias, and other examples in the hospitality sector generate approximately 12 million tonnes of FW annually [5]. However, kitchen waste of catering companies is no longer referred to as FW; instead, according to Regulation (EC) No. 1069/2009 (Animal By-products regulation) [6], the waste is now entitled as Category III animal by-products (ABPs).

Because Category III ABPs can increase the spread of contagious diseases, this waste must be managed in accordance with the health rules presented in Commission Regulation (EU) No 142/2011 [7]. According to these rules, catering companies must separate and store FW in refrigerators under specific conditions before the ABP manager transfers them for further treatment. This ABP management causes increased environmental impact due to electricity consumption for refrigerating and the frequent transportation of ABPs that contain considerable moisture content and is also expensive for a catering company [8].

Kliopova et al. (2019) indicated that costly ABP management could be avoided by adopting intensive composting equipment. Such an approach would avoid costly ABP management services and significantly reduce the environmental impact. Additionally, the resulting product was characterized by outstanding quality parameters (e.g., organic matter (OM) [85% of dry matter (DM)], water-soluble N (840 ppm), P (2089 ppm), and K (15,875 ppm)), trace amounts of heavy metals, and practically no microbiological contamination [8]. However, the primary compost contained an abundance of $SO_4^{2-}$ (2800 ppm as received) and $Cl^-$ (11,300 ppm as received) anions, which are toxic to plants at high concentrations (>300 ppm) [9]. These anions are a part of the total conductivity (5.89 mS $cm^{-1}$), which was significantly higher than the threshold limit of 4 mS $cm^{-1}$ [10]. Furthermore, the primary compost stability, expressed as the concentration of dissolved organic carbon (DOC, 87,264 mg $kg^{-1}$), was far from that of mature compost (<4000 mg $kg^{-1}$) [11]. Moreover, the sole maturation of the primary FW compost for one month did not help to stabilize the parameters (Kliopova et al., 2019). The compliance of $SO_4^{2-}$,$Cl^-$, conductivity, and biodegradability values with corresponding thresholds is crucial for safe agricultural application; therefore, additional measures to control these parameters are needed [12].

As the produced primary ABP compost is characterized by outstanding nutritive properties, it can be considered as an amendment for low-quality green waste compost (GWC). The GWC often contains minor quantities of N (0.5–1.5% DM), P (0.1–0.2% DM), K (0.4–0.8% DM) [13], and OM (between 8.5 and 28.6% DM) [9]. Poor compost quality is a problem because the majority of countries have established limits for quality compost, and if the produced compost does not comply with the limits, it is considered a waste. For instance, in Lithuania, compost produced from GW and FW must contain more than 25% DM of OM, and the sum of total N (TN), $P_2O_5$, and $K_2O$ must be over 2.5%. Similarly, the Canadian standard for certification requires compost to contain more than 0.7, 0.5, 0.75, and 20% of N, P, K, and OM, respectively, to be classified as quality compost [14]. For this reason, it is important to meet requirements to utilize products in the market.

This study investigates the possibility of improving the environmental and economic performance of catering Category III ABP management in a region by applying industrial symbiosis compared to conventional ABP management. The industrial symbiosis approach was chosen because the municipal GW stream can be incorporated into the system to optimize compost quality and increase the amount of higher value compost production. Interestingly, in this study, the GW was not exploited as a co-substrate, as typically found in other studies [15]; however, the authors suggest the use of already produced GW primary compost, which is an abundant material in municipal GW composting sites [16]. The addition of ABPs primary compost to GW primary compost would increase the concentration of important nutrients and OM in the maturated mixture, thus increasing the price of the final product. As poor nutrient concentration is considered one of the reasons for reduced compost usage, the addition would promote a more extensive application of compost in agriculture, reducing environmental impacts [17].

In general, researchers agree that FW management in the hospitality sector is scarcely documented in the scientific literature [18]. Moreover, most of the studies are focused on the management of avoidable FW [18]; these studies mostly consider preventative measures [19]. To the authors' knowledge, the combination of the management of unavoidable hospitality FW following ABP regulation, the principle of industrial symbiosis, and exploitation of existing municipal GW management systems has not been addressed by other researchers.

The main objectives of this study are as follows:

1. Conduct primary ABP compost production experiments in three catering companies, investigate quality parameters, and select one for further analysis;
2. Perform maturation experiments of selected ABPs and GW primary composts and analyze quality parameters of the produced compost;
3. Run the life cycle inventory of the currently employed ABP composting plant in the region and analyze the quality of the compost produced;
4. Compare the environmental performance of conventional management and suggest catering ABP management approaches by applying the life cycle assessment (LCA) method;
5. Run an economic assessment of the suggested alternative.

## 2. Materials and Methods

### 2.1. Experimental Conditions and Equipment

The experiment was performed at three Lithuanian catering companies that generated between 25 and 35 kg $d^{-1}$ (approx. 10 t $y^{-1}$) of Category III ABPs (FW referred to as ABPs according to Regulation (EC) No. 1069/2009). Two of them were kindergartens, which had a consistent menu such as a typical restaurant; thus, at the end of a day, leftovers and preparation waste were quite uniform because of very few different dishes. As kindergartens have a very strict diet, vegetables and grains were the prevailing ingredients for the dishes, accompanied by main entrees (meat-based) and also dairy products and fruits. Consequently, preparation and unavoidable residues (peelings of potatoes, beets, and carrots, banana skins, apples, etc.), and unconsumed meals comprise very homogeneous FW streams. Conversely, the third catering company had heterogeneous FW because meals were prepared according to the changing demands of customers. Due to differences in FW composition and its potential impact on final product parameters, the three catering companies were considered.

Across the catering companies, two different technologies for closed-loop intensive aerobic treatment were utilized: the Oklin GreenGood 10s [8] and the Green Service Bioreactor Bio-10 [20]. The three intensive automatic composting machines were installed in each company in accordance with the requirements presented in the manuals. Oklin equipment for exhaust air treatment uses an activated carbon filter, whereas Green Service applies a nano-filter deodorization unit before removing the generated gas through a ventilation system.

Both intensive composting machines are designed to treat a batch of FW in a 24-h cycle. After the cycle was completed, the primary compost produced was removed, leaving approximately 10% of the compost as an inoculum for the successive cycle. The Oklin technology incorporates a specific microorganism strain under the brand name ACIDULO™, whereas Green Service uses the Ecibulo brand, which is composed of halophilic, eosinophilic, and thermophilic germs [21]. Green Service also emphasizes the need for the inoculum and even replenishment of microorganisms every year. Microorganisms, intensive aeration, agitation, and temperature regimes are the main factors influencing rapid FW degradation, moisture evaporation, and mass reduction of up to 50–90%. During composting, Oklin equipment maintains temperatures above 55 °C, but for 1 h, the temperature is increased to 70 °C to eliminate pathogens [8]. Green Service technology follows the same pattern; during processing, the temperature is maintained in the range of 54–56 °C, but at the end of the composting, the temperature is increased to 70 °C for 1 h.

The experiment was divided into two parts. The first part was the production of primary composts in the three catering companies. The experiment started in June 2018, during which 20 cycles (one cycle duration was one working day or 24 h) were conducted in the catering companies. The first catering company transformed approximately 28 kg per cycle (total 560 kg of ABPs per 20 cycles) to 6.72 kg (total 134.3 kg of ABPs). The second catering company composted 14.63 kg per cycle (total 292.6 kg) of ABPs to 3.18 kg (total 63.5 kg). The third catering company composted 9.83 kg per cycle (total 196.5 kg) of ABPs to 2.39 kg (total 47.8 kg) of the primary compost. In each catering company, each batch (produced per cycle) was poured into a clean polypropylene bag and mixed with the other simultaneously gathered batches. After completion of the 20 cycles, the sampling procedure was performed. Three samples from the primary compost of ABPs1 were collected. Sampling was performed by taking 10 specimens from the top, middle, and bottom locations of the collected ABPs1 primary compost in random order and then uniformly mixing these 10 specimens to form one sample. The second and third samples of ABPs1 were prepared in the same manner. The same sampling procedure was applied to the primary compost of ABPs2 and ABPs3.

GW primary compost was prepared at a municipal GW composting site. The GW primary compost production started in the first week of April 2018 and ended at the end of July 2018. The waste available at the municipal GW composting site was used as a substrate. After composting was completed, three samples were taken from the pile of the primary GW compost. The samples were prepared in the same manner as the ABPs1, ABPs2, and ABPs3 samples.

The second part of the experiment was the co-maturation of the selected ABPs and GW primary composts in the municipal GW composting site. The experiment started at the beginning of August 2018. The ABPs2 primary compost was selected for the second part of the experiment because it contained higher concentrations of TN and total P (TP), and had a more attractive visual appearance than ABPs1 and ABPs3. A total of 15% of ABPs2 and 85% of GW primary composts were mixed (by mass) and loaded into a standard open home composting bin (V-approx. 700 L) for co-maturation. During maturation, the substrate was thoroughly mixed at least once a week. Finally, the samples were collected after 60 days (in October 2018) and 90 days (in November 2018) of co-maturation. All these ABPs and GW (including the compost produced in the conventional ABP treatment plant) samples were delivered to the agrochemical research and joint-stock company "Jurby Water Tech" laboratories to analyze the quality parameters.

### 2.2. Evaluation of Mass Balance during Composting Processes

The Utena municipality (Lithuania) was chosen to assess current catering ABP management and compare it to the new approach. The most important features of the region were the number and location of municipal GW composting sites, distance from a catering company to ABP managers, and technology used for ABP management. The average distance from the catering company to the municipal GW site was 15 km, whereas the distance between the companies and ABPs managers was 50 km.

The N balance was calculated according to the study by Yang et al. (2019) [22]. These authors found that the N loss via $N_2O$ was 0.07%, N loss via $NH_3$ was 9.88%, N loss via leachate was 0.38%, unaccounted N was 2.81%, and TN loss was 13.13% from the initial N content for forced aeration systems. Similarly, N balance was assessed for passive aeration systems [22]. In this case, the N loss via $N_2O$ was 0.09%, N loss via $NH_3$ was 8.10%, N loss via leachate was 5.17%, unaccounted N was 2.62%, and TN loss was 15.98% of the initial N content.

The P balance was evaluated using the study of Iqbal et al. (2015). These authors found that the TP loss with filtrate during composting was approximately 3% from initial P [23]. Similarly, based on the study of Sommer (2001), the loss of total K (TK) with filtrate was 12% from initial K [24]. CO emissions were assessed according to EMEP/CORINAIR, whereas $CH_4$ emissions for open (3.4 kg $t^{-1}$ of waste) and closed (0.9 kg $t^{-1}$ of waste) composting were taken from the study of Boldrin et al. (2009) [25]. The heavy metal loss was calculated considering that 2% of the heavy metals present in compost will be washed with leachate [26].

### 2.3. Life Cycle Assessment Methodology

The environmental impacts of the conventional and suggested catering ABP management scenarios were calculated and compared using the LCA method. The ReCiPe method at the midpoint level was used to perform the impact assessment based on ReCiPe2008 by Goedkoop et al. (2008) [27] and an updated version ReCiPe2016 by the National Institute for Public Health and the Environment [28]. In terms of the environmental impact categories at the midpoint, all 18 impact categories were calculated. The LCA Ecoinvent Database v3.6. (2019) was applied as the background source for life cycle impact analysis [29]. The potential environmental impacts of management scenarios were calculated using the LCA software SimaPro 9.1 [30]. The functional unit used was 10 t $y^{-1}$ of the ABPs.

## 3. Results and Discussion

### 3.1. Parameters of Primary Composts

The analysis of the primary composts obtained and produced during the first part of the experiment showed expected yet variable quality parameters (see Table 1). The analysis of the produced ABPs1, ABPs2, and ABPs3 primary composts confirmed that the quality parameters rely heavily on the input FW. For instance, ABPs2 and ABPs3 from kindergartens had relatively similar menus; hence, the produced primary composts were similar as well. However, noticeable differences in NPK values between ABPs2 and ABPs3 kindergarten primary composts were observed. Regarding TN, the ABPs2 primary compost contained 2.38% DM, whereas ABPs3 had slightly less (1.92% DM), showing that ABPs2 had more N-rich waste such as fish or meat meal residues. The ammonification rate difference between the primary compost of the ABPs1 (518 ppm $NH_4^+$) and ABPs2 (1534 ppm $NH_4^+$) was notable, but these two catering companies utilized separate intensive composting technologies with diverse microbial communities. However, it is possible that the inoculation procedure was disrupted, resulting in highly fluctuating (518–1534 ppm) $NH_4^+$ concentrations in the primary composts analyzed. Regarding the inoculum, technology applied for the ABPs2 and ABPs3 treatments used halophiles, which are effective under extreme conditions such as low pH (<4), high conductivity (>5 mS $cm^{-1}$), and temperature (>55 °C), potentially influencing the swift mineralization of biomass [31].

Because these extreme conditions are detrimental to typical composting microflora, these microbes can easily dominate. Moreover, when the temperature was increased to 70 °C, pathogens and potentially the most typical composting microbes were eliminated. For this reason, when processed, ABPs were removed from the automatic composting machine and left at ambient temperature, and the microbial activity was significantly reduced. As a result, the ammonification process was impeded, leading to the preservation of $NH_4^+$ in the compost. In addition to high temperature, forced air supply for 24 h of batch processing also influenced the N behavior. Most of the time, the temperature was in the range of 40–60 °C, which is optimum for conversion or organic N to ammonium and inhibitory for nitrifying bacteria [32]. Therefore, $NH_4^+$ dominated $NO_3^-$ in the primary compost. The forced air supply may have also contributed to the $NH_4^+$ concentration. According to de Guardia et al. (2010), the aeration rate can considerably increase the ammonification rate, for example, from 36.7 to 43.1% for household waste or from 52.0 to 71.4% for pig slaughterhouse sludge [33].

The pH of ABPs1, ABPs2, and ABPs3 primary composts was correlated with $NH_4^+$ concentration. Because $NH_4^+$ increases the pH, the highest pH was found in the specimen containing the largest amount of $NH_4^+$ (1534 ppm). As $NH_4^+$ concentrations decreased in ABPs3 (970 ppm) and ABPs1 (518 ppm) specimens, the pH decreased to 5.0 and 4.7, respectively.

**Table 1.** Major agronomical parameters of primary GW and ABPs composts.

| Quality Indicators * | Primary GW Compost | Primary ABPs1 Compost | Primary ABPs2 Compost | Primary ABPs3 Compost |
|---|---|---|---|---|
| *E. coli* **, cfu $g^{-1}$ | $\leq 1 \times 10^3$ | $\leq 1 \times 10^3$ | $\leq 1 \times 10^3$ | $\leq 1 \times 10^3$ |
| Conductivity, mS $cm^{-1}$ | **1.04** (1.02–1.08) | **5.77** (5.70–5.81) | **5.40** (5.37–5.42) | **5.51** (5.47–5.55) |
| $pH_{KCl}$ | **7.4** (7.3–7.5) | **4.70** (4.6–4.8) | **7.1** (7.0–7.1) | **5.0** (4.9–5.1) |
| DM % | **66.18** (64.13–67.21) | **88.92** (86.97–91.20) | **84.21** (82.08–87.04) | **75.77** (72.16–78.21) |
| OM, % DM | **18.31** (16.94–19.23) | **83.22** (82.14–84.63) | **88.34** (87.48–89.35) | **92.89** (91.94–93.85) |
| TN, % DM | **0.72** (0.69–0.75) | **1.81** (1.79–1.84) | **2.38** (2.35–2.41) | **1.92** (1.86–1.98) |
| TP, % DM | **0.71** (0.69–0.75) | **0.39** (0.38–0.40) | **0.65** (0.60–0.67) | **0.54** (0.49–0.53) |
| TK, % DM | **0.62** (0.59–0.64) | **1.44** (1.40–1.51) | **1.26** (1.23–1.28) | **1.02** (1.01–1.04) |
| Water-soluble N, ppm | **189** (182–195) | **546** (520–566) | **1534** (1501–1563) | **1015** (967–1045) |
| Water-soluble $NO_3^-$ + $NO_2^-$, ppm | **159** (153–164) | **27.82** (25.03–31.15) | **52.89** (48.21–58.01) | **45.05** (40.14–47.67) |
| Water-soluble $NH_4^+$, ppm | **30.39** (26.71–35.54) | **518** (489–524) | **1481** (1453–1505) | **970** (919–998) |
| Water-soluble P, ppm | **53.6** (51.0–55.2) | **1321** (1302–1349) | **1898** (1845–1915) | **1917** (1851–1966) |
| Water-soluble K, ppm | **854** (812–887) | **10,319** (10,204–10,495) | **10,745** (10,406–11,008) | **11,054** (10,524–11,386) |
| Cl, ppm | **471** (428–511) | **7155** (7120–7230) | **9240** (9028–9484) | **9797** (9715–9887) |
| $SO_4$, ppm | **628** (598–662) | **1820** (1773–1858) | **1089** (1026–1184) | **987** (976–1003) |
| DOC, mg $kg^{-1}$ | **486** (446–517) | **80,892** (80,506–81,508) | **42,970** (42,397–43,394) | **41,190** (41,024–41,386) |
| Bulk density, g $L^{-1}$ | 795 | 688 | 776 | |

Notes: * Data are reported as mean with n = 3 and the largest and lowest values in parentheses. **—Contamination by *E. coli* was measured only once per sample.

The conductivity among the primary composts varied from 5.4 to 5.77 mS $cm^{-1}$, possibly contributing to extreme conditions. For example, Setia et al. (2010) stated that at a conductivity above 5.0 mS $cm^{-1}$, soil respiration declined over 50% [34], whereas other researchers reported that the inhibitory value for microbial activity during composting was 8 mS $cm^{-1}$ [35]. In addition to inhibitory effects, the high conductivity of the primary composts illustrates the effective release of $NH_4^+$, P, and K from organic compounds. These conductivity values were far from those of composts used in agriculture; for instance, Turan (2008) stated that 2 mS $cm^{-1}$ is the ideal value [36], whereas Staugaitis et al. (2016) referred to 1.1–1.5 mS $cm^{-1}$ as optimum values for stable products.

The unfinished degradation process is clearly illustrated by the abundance of DOC, also referred to as biodegradability (compost stability). The ABPs2 (42,970 mg $kg^{-1}$) and ABPs3 (41,190 mg $kg^{-1}$) primary composts had nearly identical concentrations of DOC, whereas that of ABPs1 was twice as high (80,892 mg $kg^{-1}$). All of these primary composts exceeded the stable compost threshold of <4000 mg $kg^{-1}$ by 10 to 20 times [11]. Although the primary compost was not stable, high concentrations of DOC indicated microbial activity [37]. In addition, Kliopova et al. (2019) reported that primary compost produced in an intensive composting machine had relatively high concentrations of fulvic and humic acids, which indicates that the composting process was not simple evaporation of water [8].

Water-soluble P solubilization during the treatment period was comparatively high. The soluble P-concentration ranged from 1321 to 1917 ppm (1920–2477 mg $kg^{-1}$ according to the bulk density). The high solubilization rate may be due to low molecular weight acids in the total pool of DOC and the activity of P-solubilizing microorganisms [38].

The concentration of water-soluble K accumulated for 24 h was extremely high. The water-soluble P was nearly identical in samples from the catering companies and varied from 10,319 to 11,054 mg kg−1. K is not incorporated in organic compounds such as N or P; instead, it is concentrated in cells in the form of salts [39]. For this reason, when cells collapse during composting, K salts are released into the media, resulting in a high K concentration.

In the analyzed samples, contamination by *E. coli* was below the limit value (see Table 1). These results are consistent with those of a previous study by Kliopova et al. (2019) and Pandey et al. (2016), who also indicated that closed-loop composting equipment can produce compost without any pathogenic microorganisms [40].

The GW primary compost had typical values found across Lithuania municipal GW composting sites [9]. Generally, the primary compost had low quantities of TN, TP, and TK, as well as their soluble forms. In contrast to catering companies, in primary GWC, the nitrates were dominant over $NH_4^+$, indicating the prevalence of nitrifying bacteria.

### 3.2. Parameters of Cured Compost

The co-maturation of ABPs2 and GW primary composts resulted in a product with no phytotoxic qualities (see Table 2). The conductivity (1.1 mS $cm^{-1}$) was in the optimum range, excluding the possibility of osmotic imbalance (dehydration) of a plant cell. The pH after 60 days of maturation was 6.63, but reached 7.0 after 90 days. This is the optimal pH because N, K, Ca, Mg, and S are more available in the pH range of 6.5–8.0, whereas P is more available in a pH between 5.5 and 7.5 [41]. The DOC (82 mg $kg^{-1}$) was significantly lower than the threshold (<4000 mg $kg^{-1}$), resulting in a very stable product. In fact, even if the co-maturation duration was shortened, the compost stability would have met the requirement. $Cl^-$ concentration (298 ppm), which is another parameter associated with toxicity in plants, was also below the limit value of > 300 ppm. The $SO_4^{2-}$ concentration (808 ppm) remained high, and concentrations above 300 may be detrimental for susceptible plants [9]. However-, the concentration of sulfate (128 ppm) was reduced significantly after 90 days.

**Table 2.** Results of co-maturation of GW and ABPs primary composts.

| Parameters for Quality Indication | ** Values for Quality Indication | * Maturated Composts ABPs2 + GW | | Quality Assessment of Compost (After 2 Months) |
|---|---|---|---|---|
| | | After 2 Months | After 3 Months | |
| Conductivity, mS $cm^{-1}$ | <0.6 → > 2 | **1.1** (1.07–1.14) | **0.9** (0.98–0.92) | Medium |
| $pH_{KCl}$ | <5.6 → 8.5 | **6.63** (6.56–6.71) | **7.0** (6.9–7.1) | Medium |
| DM, % | <21 →>50 | **58.18** (56.89–59.6) | **60.35** (60.05–60.60) | Very high |
| OM, % DM | <16 → >45 | **20.32** (19.81–21.17) | **17.30** (17.01–17.82) | Low |
| N, % DM | <0.5 → >2.0 | **1.36** (1.28–1.42) | **0.97** (0.89–1.08) | Medium |
| P, % DM | <0.21 → >0.8 | **0.93** (0.80–1.01) | **0.76** (0.69–0.82) | Very high |
| K, % DM | <0.6 → >2.5 | **0.76** (0.70–0.82) | **0.45** (0.42–0.47) | Low |
| Water-soluble N, ppm | <51 → >200 | **388** (372.0–406.1) | **344** (341–348) | Very high |
| Water-soluble $NO_3^-$ + $NO_2^-$, ppm | <51 → >200 | **387** (371–405) | **344** (341–348) | Very high |

**Table 2.** *Cont.*

| Parameters for Quality Indication | ** Values for Quality Indication | * Maturated Composts ABPs2 + GW | | Quality Assessment of Compost (After 2 Months) |
|---|---|---|---|---|
| | | After 2 Months | After 3 Months | |
| Water-soluble $NH_4^+$, ppm | - | **1.06** (1.03–1.10) | **0.013** (0.010–0.018) | - |
| Water-soluble P, ppm | <26 → >100 | **214** (236–250) | **190** (188–192) | Very high |
| Water-soluble K, ppm | <91 → >300 | **832** (809–862) | **700** (697–703) | Very high |
| Water-soluble Ca, ppm | <101 → >500 | **318** (311–325) | **201** (196–205) | High |
| Water-soluble Mg, ppm | <31 → >120 | **105** (95–114) | **81** (80–83) | High |
| Cl, ppm | <51 → > 300 | **298** (271–328) | **248** (241–258) | High |
| $SO_4$, ppm | <51 → > 200 | **808** (788–832) | **123** (119–128) | Very high |
| DOC, mg $kg^{-1}$ | < 4000 | 82 | 0 | Stable compost |
| *E. coli*, cfu $g^{-1}$ | $\leq 1\times10^3$ | $6.4\times10^4$ | $2.3\times10^3$ | Polluted |

Notes: * Data are reported as mean with n = 3 and the largest and lowest values in parentheses. **—Maturated compost quality was assessed according to Staugaitis et al. (2016). The first and second values refer to the lower (very low quality) and upper (very high quality) limits of the parameter in question.

Another parameter for compost stability determination is the $NH_4^+$ to $NO_3^-$ ratio. Many researchers claim that the $NH_4^+$ to $NO_3^-$ ratio should be less than 1. In the experiment, the $NH_4^+$:$NO_3^-$ ratio was only 0.0027, suggesting excellent stability. Moreover, due to effective $NH_4^+$ oxidation, the maturation period could be considerably reduced, maintaining the $NH_4^+$ to $NO_3^-$ ratio lower than one.

Many parameters associated with the fertility of the product exceeded the value, which was determined to be valuable according to a study conducted by Staugaitis et al. (2016). The compost produced in the experiment had several parameters that surpassed a very high rating, such as TP (0.93%; very high > 0.8%); water-soluble N (388 ppm; very high > 200 ppm); water-soluble P (214 ppm; very high > 100 ppm); water-soluble K (808 ppm; very high > 300 ppm); DM (58.18%; very high > 50%) (see Table 2). However, the most important agronomical parameter of OM was considerably low because of the prevalence of low OM containing GW primary compost and degradation of organic material during maturation. Nevertheless, if the ABPs2 portion in the mixture was increased by 20 to 30%, the OM percentage would notably increase.

Although the parameters associated with phytotoxicity and compost fertility exhibited excellent values, during the experiment, *E. coli* ($6.4 \times 10^4$) exceeded the limit value of $1 \times 10^3$ cfu $g^{-1}$, and the microbial count of the GW and ABPs primary composts was below the threshold value. The increase in *E. coli* colony count could be due to uncontrolled disturbances and low temperatures during maturation.

As a solution for emerging *E. coli* problems, specific microorganism mixtures can be introduced during co-maturation. After the analysis conducted in the Department of Veterinary Medicine, Jelgava University of Agriculture, the specifically developed probiotic mixture (microbiological preparations) showed that *E. coli* (strain No. ATCC25922) colony numbers were significantly reduced in only 24 h [42]. The comparison of both produced composts revealed that composting with the probiotics completely eliminated *E. coli* and Salmonella as well as other types of pathogens (e.g., *Enterobacter cloacae*) [43].

### 3.3. Current and Planned Hospitality ABPs Management Approach

To reveal the environmental benefits of catering ABP management, a comparison of existing and suggested approaches to treat 10 tons of ABPs was performed. First, a field study of the current ABP plant was conducted. During the study, the researchers found that the technology used was composting under fabric cover with forced aeration (air supply through perforated tubes under the piles) to minimize the need for pile turning (a windrow turner was used only two times per composting cycle) and accelerate the composting process. However, compost turning was applied for mixing catering ABPs (55%), GW (35%), and biomass combustion ash (10%) to adjust the moisture content and C/N ratio. After surveying the plant's staff, it was assumed that the substrate mass was reduced by approximately 62%, whereas the dry material reduction was approximately 36%. Composting with forced aeration generated 6.07 t $y^{-1}$ of leachate.

After the plant staff survey and analysis of the produced compost quality and contamination parameters (average value n = 3), and analysis of scientific literature, the mass balance was calculated (see Figure 1). According to the compost quality analysis and loss of material during composting (see Section 2.2), the NPK in the input material was calculated.

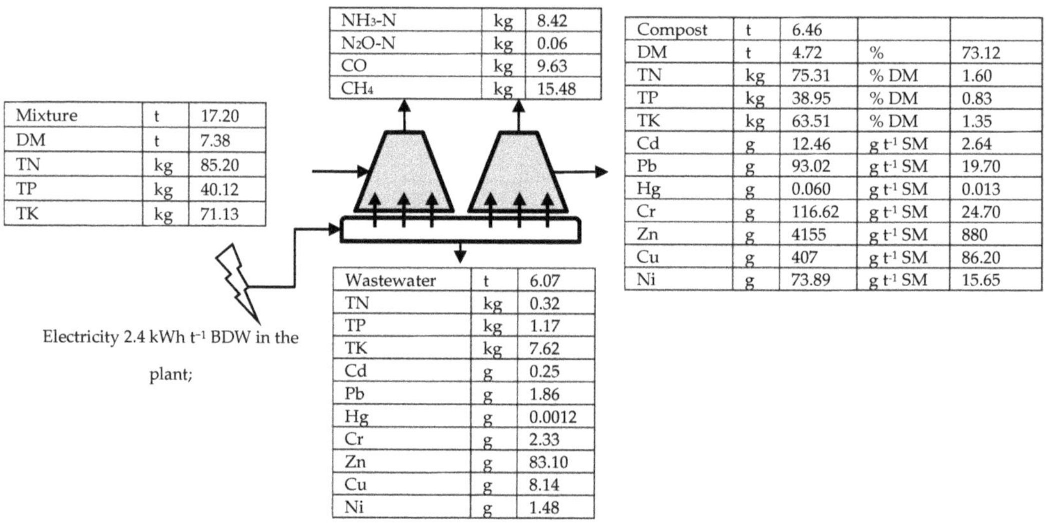


| Mixture | t | 17.20 |
|---|---|---|
| DM | t | 7.38 |
| TN | kg | 85.20 |
| TP | kg | 40.12 |
| TK | kg | 71.13 |

| $NH_3$-N | kg | 8.42 |
|---|---|---|
| $N_2O$-N | kg | 0.06 |
| CO | kg | 9.63 |
| $CH_4$ | kg | 15.48 |

| Compost | t | 6.46 | | |
|---|---|---|---|---|
| DM | t | 4.72 | % | 73.12 |
| TN | kg | 75.31 | % DM | 1.60 |
| TP | kg | 38.95 | % DM | 0.83 |
| TK | kg | 63.51 | % DM | 1.35 |
| Cd | g | 12.46 | g $t^{-1}$ SM | 2.64 |
| Pb | g | 93.02 | g $t^{-1}$ SM | 19.70 |
| Hg | g | 0.060 | g $t^{-1}$ SM | 0.013 |
| Cr | g | 116.62 | g $t^{-1}$ SM | 24.70 |
| Zn | g | 4155 | g $t^{-1}$ SM | 880 |
| Cu | g | 407 | g $t^{-1}$ SM | 86.20 |
| Ni | g | 73.89 | g $t^{-1}$ SM | 15.65 |

| Wastewater | t | 6.07 |
|---|---|---|
| TN | kg | 0.32 |
| TP | kg | 1.17 |
| TK | kg | 7.62 |
| Cd | g | 0.25 |
| Pb | g | 1.86 |
| Hg | g | 0.0012 |
| Cr | g | 2.33 |
| Zn | g | 83.10 |
| Cu | g | 8.14 |
| Ni | g | 1.48 |

**Figure 1.** Existing hospitality ABPs management in the selected region.

After material flow analysis, the energy demand data for the composting operations were gathered. Electricity consumption in the plant was 2.4 kW h $t^{-1}$ of biodegradable waste (BDW), amounting to a total of 41.29 kWh for 17.2 t $y^{-1}$ of the substrate. Regarding diesel consumption, the staff reported that all the machinery consumed 1.26 kg t $y^{-1}$ BDW, resulting in a total 21.68 kg of fuel for the composting operation.

Conventional catering ABP management in the selected region requires considerable energy for transportation. In the case of the catering company, which generates 10 t $y^{-1}$ ABPs, it is assumed that the ABPs will be collected every week (52 picks per year), consisting of 192.3 kg per pick. Because the average distance between the catering company and the ABP manager is 50 km, the need for transportation is 26,000 tkm y−1. Moreover, the transportation of the product (6.46 t $y^{-1}$) to agricultural fields in the region would also amount to 50 km, amounting to a total of 306.74 tkm.

The suggested management approach is shown in Figure 2. The flowchart presents the management of one catering company's ABPs by applying intensive composting at the source of generation, followed by co-maturation of GW and ABPs primary composts in the closest municipal GW composting sites.

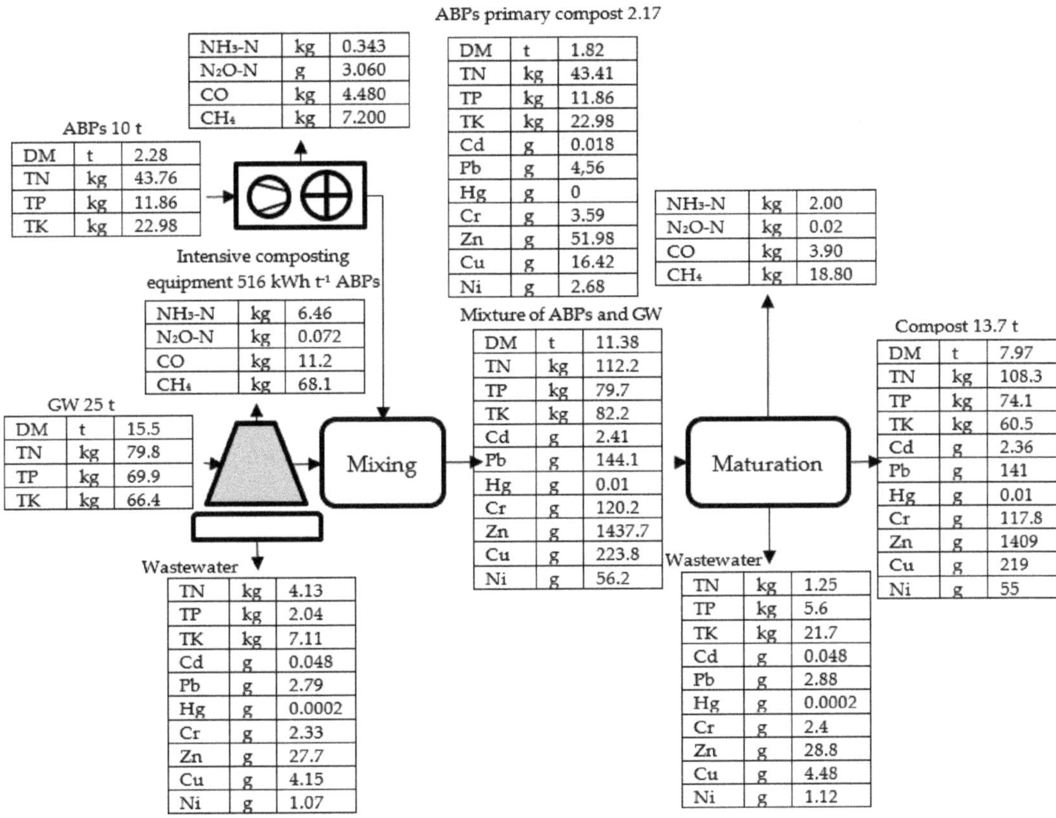


**Figure 2.** Existing hospitality ABPs management in the selected region.

The scheme in Figure 2 presents the flow of nutrients and materials during the application of the novel management approach. The box with an air blower and agitator presets intensive composting on the premises of a catering company, and the gray pile depicts typical municipal GW composting.

During the intensive composting of catering ABPs, no wastewater was generated due to external heating and evaporation. For this reason, the quantities of K and P remained the same in the lower-mass material. In a previous study, Kliopova et al. (2019) reported that 0.0416 kg of $NH_3$ was generated per ton of catering ABPs. These findings were used to evaluate the loss of TN in the form of $NH_3$ (0.343 kg) from the system. As $N_2O$ emissions were not measured, the emissions factor for aerated compost was taken from Yang et al. (2019). The N balance was calculated in accordance with N emissions.

Concentrations of heavy metals were also taken from the study of Kliopova et al. (2019), as additional laboratory analysis would be redundant due to the nature of source segregated FW, which is nearly heavy metals-absent. Heavy metal concentrations of GWC were taken from Staugaitis et al. (2016) as a more reliable source of data. The authors analyzed composts of multiple GW composting sites and showed average heavy metal concentrations across Lithuania. The heavy metal concentrations in both primary composts were used to assess the amount of heavy metals in the mixture before maturation.

$CH_4$ and CO emissions were distributed across the composting and maturation processes. A total of 80% of $CH_4$ and CO were assigned to composting and 20% to maturation.

During the novel ABP management, the majority of diesel is consumed during municipal GW composting. The staff of the GW composting plant of the region reported 0.756 kg per ton of GW, representing 18.9 kg per 25.0 t for GW composting. In the case of the suggested management approach, diesel is used to mix ABPs and GW primary composts before maturation by using the plant compost-turning machine (capacity 2800 $m^3$ $h^{-1}$) with diesel consumption of 16 L $h^{-1}$ or 0.016 L per ton of BDW (one cycle). It is assessed that two cycles would be sufficient to ensure proper mixing of the primary composts, amounting to 0.23 kg of diesel for 16.6 tons of primary composts.

Most importantly, the novel ABP management approach proposed can substantially reduce transportation needs. As the production of primary compost in a catering company is no longer waste, ABP regulation does not apply to this subproduct. Therefore, the transport of the primary compost to the closest municipal GW composting sites is suggested. In the case of the selected region, 10 t $y^{-1}$ of ABPs would be transformed into 2.17 t $y^{-1}$ of the primary compost and transported for 15 km (empty return) to the closest municipal GW composting site (32.49 tkm) every two weeks or 26 times per year (83.3 kg). Moreover, the delivery of the produced compost to the surrounding agricultural fields would be less intensive compared to the current management approach. Six GW composting sites with a radius of 15 km cover 85% of the region's territory. For this reason, the delivery of the product (10.29 t $y^{-1}$) for field application would amount to 154.16 tkm.

### 3.4. Current and Planned Hospitality ABPs Management Environmental Impact Assessment

To highlight the environmental benefits, current and planned catering ABP management approaches were interpreted by applying the LCA method. After the analysis (characterization) of both approaches, a reduction was observed in 12 environmental impact categories from a total of 18. For instance, the environmental impact in the climate change category was reduced by 48%, ozone formation by 107%, fossil resource scarcity by 83%, etc. (see Figure 3). The same impact categories with their corresponding units were depicted in Table 3.

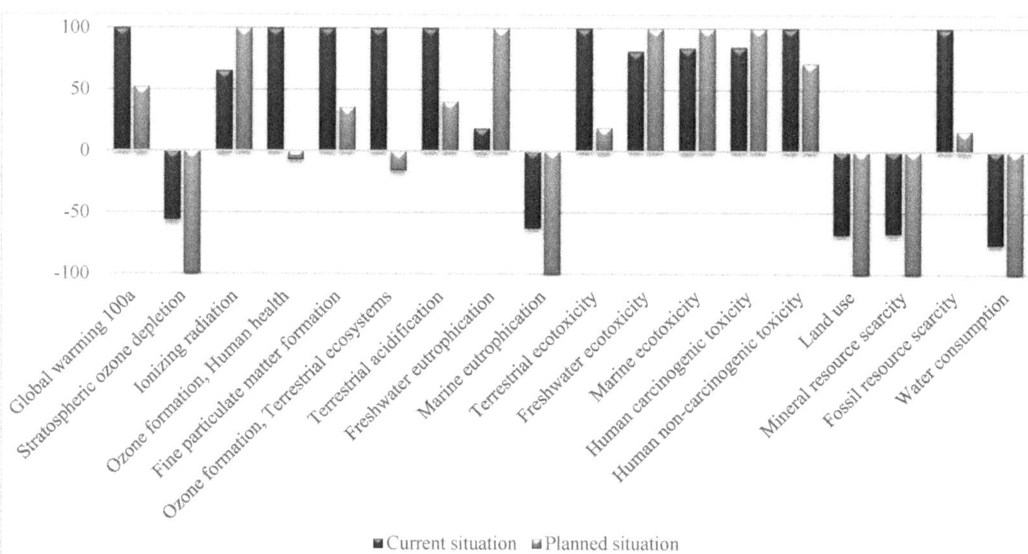


**Figure 3.** Comparison of environmental impacts according to the LCA method for current and planned catering ABP management approaches.

**Table 3.** Comparison of current and planned situations in 18 environmental impact categories.

| | Impact Category | Unit | Current Situation | Planned Situation |
|---|---|---|---|---|
| 1 | Global warming 100a | kg $CO_2$ eq | 7935 | 4141 |
| 2 | Stratospheric ozone depletion | kg CFC11 eq | −0.0087 | −0.0156 |
| 3 | Ionizing radiation | kBq Co-60 eq | 646 | 988 |
| 4 | Ozone formation, Human health | kg NOx eq | 80.9 | −5.48 |
| 5 | Fine particulate matter formation | kg PM2.5 eq | 14.8 | 5.32 |
| 6 | Ozone formation, Terrestrial ecosystems | kg NOx eq | 77.0 | −11.8 |
| 7 | Terrestrial acidification | kg $SO_2$ eq | 57.7 | 23.6 |
| 8 | Freshwater eutrophication | kg P eq | 1.57 | 8.28 |
| 9 | Marine eutrophication | kg N eq | −0.40 | −0.64 |
| 10 | Terrestrial ecotoxicity | kg 1,4-DCB | 18,236 | 3572 |
| 11 | Freshwater ecotoxicity | kg 1,4-DCB | 173 | 212 |
| 12 | Marine ecotoxicity | kg 1,4-DCB | 221 | 263 |
| 13 | Human carcinogenic toxicity | kg 1,4-DCB | 63.7 | 75.2 |
| 14 | Human non-carcinogenic toxicity | kg 1,4-DCB | 2969 | 2129 |
| 15 | Land use | $m^2$a crop eq | −4,567,849 | −6,760,311 |
| 16 | Mineral resource scarcity | kg Cu eq | −7.16 | −10.7 |
| 17 | Fossil resource scarcity | kg oil eq | 2364 | 413 |
| 18 | Water consumption | $m^3$ | −34.9 | −46.2 |

### *3.5. Economic Assessment of the Proposed Management Approach*

The savings for a catering company were evaluated in a previous study (Kliopova et al., 2019), amounting to 4203 EUR per year, when the produced compost is used in the catering company premises. However, when the primary compost cannot be matured and used for the needs of the company, it is suggested to divert the primary compost to GW composting sites, as it was analyzed in the present study. In this case, system boundaries are enlarged, encompassing municipal GW composting sites, which also receive extra earnings. The pay-pack period of implementation of an intensive composter will be approximately 2.1 years (see Table 4).

**Table 4.** Economic evaluation (savings) due to installation of an intensive composter at the catering company.

| Analyzed Parameters | Units | Price | Situation before Implementation (Conventional ABPs Management) | Situation after Implantation of Intensive Composter | Savings | |
|---|---|---|---|---|---|---|
| | | EUR unit$^{-1}$ | Units y$^{-1}$ | Units y$^{-1}$ | Unites y$^{-1}$ | EUR y$^{-1}$ |
| ABPs | t | 446.11 | 10 | 0 | 10 | 4461.1 |
| PC | t | - | 0 | 2.17 | −2.17 | 0 |
| * Electricity | kWh | 0.15 | 3300 | 5160 | −1860 | −279.00 |
| ** Diesel fuel | L | 1.00 | 0 | 54.6 | −54.6 | −54.6 |
| | | | Total savings: | | | 4127.5 |

Notes: * Animal by-products in a catering company are stored in a separate freezer with an installed power of 0.35 kW); electricity consumption of intensive composter: 5160 kWh t$^{-1}$ of raw ABPs. ** Diesel fuel consumption for ABPs primary compost transportation to GW composting site: 32.49 tkm (diesel consumption 7 L km$^{-1}$).

As it was suggested that ABP management increases the nutritive properties of GW compost, the price of the improved compost is assumed to increase from 7–13 to 20 EUR t$^{-1}$. For this reason, established symbiotic relationships between catering and municipal GW composting sites would allow them to double the income as GW compost producers. In the case of our example, 13.7 t of higher quality compost can be produced using 2.17 t of ABPs primary compost as an amendment; expected incomes were 274 EUR or on average 137 EUR more in comparison to the current situation per catering company.

## 4. Conclusions

The analysis of ABP primary compost produced by the three catering companies has shown outstanding nutritive properties, such as the amount of OM (75–89% DM), TN content (1.5–1.38% DM), and water-soluble P concentration (1321–1917 ppm). However, the primary composts were phytotoxic, containing large concentrations of $SO_4^{2-}$ (987–1820 ppm), $Cl^-$ (7155–9797 ppm), high conductivity (5.40–5.77 mS $cm^{-1}$), and biodegradability (41,190–80,892 mg $kg^{-1}$).

The co-maturation experiment with ABPs2 and GW primary compost resulted in the stabilization of parameters associated with phytotoxicity. The cured compost conductivity was in the optimal range (1.1 mS $cm^{-1}$), biodegradability (82 mg $kg^{-1}$) in a stable range, $Cl^-$ (298 ppm) below the limit considered detrimental for the susceptible plant; however, $SO_4^{2-}$ remained high 808 ppm. According to the selected quality assessment method, the quality of the cured compost was exceptional. Parameters such as DM, TP, and soluble forms of NPK in the produced compost ranged from high to very high values.

The analysis of the current ABP management approach in the analyzed region showed that these catering wastes were composted by centralized intensive composting under forced aeration and fabric cover. In the plant, ABPs (55%) were mixed with GW (35%) and biomass combustion ash (10 %) to adjust the moisture content and C/N ratio. In addition, to treat 10 t $y^{-1}$ of ABPs, 21.68 kg of diesel was needed. In contrast, the transportation need for the current approach amounts to 26,000 km. Although the nutritive properties of the compost were high, the compost contained large amounts of heavy metals (e.g., 880 g Zn $t^{-1}$ DM).

The analysis of the environmental impacts of current and suggested ABP management approaches has revealed that in the case of implementation of the novel method, environmental performance can be increased in 12 impact categories from a total of 18. Marked reductions were achieved in climate change (48%), ozone formation (107%), and fossil resource scarcity (83%) impact categories.

The economic assessment proved that the symbiotic relationship between catering companies and GW composting sites results in economic benefits for the involved parties. Production of primary compost from 10 t of ABPs and delivery to the closest GW composting site would allow savings of EUR 4.1 thousand per year for a catering company. Additionally, co-maturation of GW and ABPs primary compost would allow a production of 13.7 tons of higher value compost in the composting site, which is two times higher than typical GW compost in the Lithuanian market because of better nutritional properties.

**Author Contributions:** Conceptualization, E.S. and I.K.; methodology, E.S. and D.K.; software, D.K.; validation, E.S., I.K. and R.P.B.; formal analysis, E.S.; investigation, E.S. and I.K; resources, E.S.; data curation, E.S.; writing—original draft preparation, E.S., I.K. and R.P.B.; writing—review and editing, I.K. and D.K; visualization, E.S.; supervision, I.K. All authors have read and agreed to the published version of the manuscript.

**Funding:** This research received no external funding.

**Institutional Review Board Statement:** Not applicable.

**Informed Consent Statement:** Not applicable.

**Conflicts of Interest:** The authors declare no conflict of interest.

## References

1. FAO. *How to Feed the World in 2050*; High-Level Expert Forum; FAO: Rome, Italy, 2009.
2. Gustavsson, J.; Cederberg, C.; Sonesson, U. *Global Food Losses and Food Waste*; Save Food Interpack Düsseldorf Ger: Quebec City, QC, Canada, 2011.
3. Hegnsholt, E.; Unnikrishnan, S.; Pollmann-Larsen, M.; Askelsdottir, B.; Gerard, M. Tackling the 1.6-Billion-Ton Food Loss and Waste Crisis. Available online: https://www.bcg.com/publications/2018/tackling-1.6-billion-ton-food-loss-and-waste-crisis.aspx (accessed on 17 December 2018).
4. Kelleher, M.; Robins, J. What Is Waste Food? *Biocycle* **2013**, *54*, 36.

5. Oliveira, B.; De Moura, A.P.; Cunha, L.M. Reducing Food Waste in the Food Service Sector as a Way to Promote Public Health and Environmental Sustainability. In *Climate Change And Health: Improving Resilience And Reducing Risks*; Filho, L., Azeiteiro, W., Alves, U.M., Sud, F., Eds.; Springer International Publishing: Cham, Germany, 2016; pp. 117–132. [CrossRef]
6. Regulation (EC) No. 1069/2009. Regulation Ec No 1069/2009 of the European Parliament and of the Council of 21 October 2009 Laying Down Health Rules as Regards Animal by-Products and Derived Products not Intended for Human Consumption and Repealing Regulation (Ec) No 1774/2002. Available online: https://eur-lex.europa.eu/eli/reg/2009/1069/oj (accessed on 7 December 2018).
7. Commission Regulation (Eu) No 142/2011. Commission Regulation (Eu) No 142/2011 of 25 February 2011 Implementing Regulation (EC) No 1069/2009 of the European Parliament and of the Council. Available online: https://eur-lex.europa.eu/legal-content/EN/ALL/?uri=CELEX%3A32011R0142 (accessed on 7 December 2018).
8. Kliopova, I.; Staniškis, J.K.; Stunžėnas, E.; Jurovickaja, E. Bio-nutrient recycling with a novel integrated biodegradable waste management system for catering companies. *J. Clean. Prod.* **2018**, *209*, 116–125. [CrossRef]
9. Staugaitis, G.; Kliopova, I.; Mažeika, R.; Gvildienė, K.; Jurovickaja, E. Arrangement of Requirements (Criterions) for Products Made of Biodegradable Waste. Unpublished work. 2016.
10. Meharg, A. Marschner's Mineral Nutrition of Higher Plants. 3rd Edition. Edited by P. Marschner. Amsterdam, Netherlands: Elsevier/Academic Press (2011), pp. 684 ISBN 978-0-12-384905-2. *Exp. Agric.* **2012**, *48*, 305. [CrossRef]
11. Zmora-Nahum, S.; Markovitch, O.; Tarchitzky, J.; Chen, Y. Dissolved organic carbon (doc) as a parameter of compost maturity. *Soil Biol. Biochem.* **2005**, *37*, 2109–2116. [CrossRef]
12. Zhang, L.; Sun, X. Evaluation of maifanite and silage as amendments for green waste composting. *Waste Manag.* **2018**, *77*, 435–446. [CrossRef]
13. Reyes-Torres, M.; Oviedo-Ocaña, E.R.; Dominguez, I.; Komilis, D.; Sánchez, A. A systematic review on the composting of green waste: Feedstock quality and optimization strategies. *Waste Manag.* **2018**, *77*, 486–499. [CrossRef]
14. Langenberg, W.J. A "Fresh" Look at Manure-"Compost"-an Iimportant "Added-Value Horticultural Product" Setting the Standards for Certification. Available online: http://www.compost.org/pdf/certifica.pdf (accessed on 4 November 2018).
15. Neugebauer, M.; Sołowiej, P. The use of green waste to overcome the difficulty in small-scale composting of organic household waste. *J. Clean. Prod.* **2017**, *156*, 865–875. [CrossRef]
16. Cao, L.; Luo, G.; Tsang, D.C.; Chen, H.; Zhang, S.; Chen, J. A novel process for obtaining high quality cellulose acetate from green landscaping waste. *J. Clean. Prod.* **2018**, *176*, 338–347. [CrossRef]
17. Ayilara, M.S.; Olanrewaju, O.S.; Babalola, O.O.; Odeyemi, O. Waste management through composting: Challenges and potentials. *Sustainability* **2020**, *12*, 4456. [CrossRef]
18. Pirani, S.I.; Arafat, H.A. Solid waste management in the hospitality industry: A review. *J. Environ. Manag.* **2014**, *146*, 320–336. [CrossRef]
19. Filimonau, V.; Todorova, E.; Mzembe, A.; Sauer, L.; Yankholmes, A. A comparative study of food waste management in full service restaurants of the united kingdom and the netherlands. *J. Clean. Prod.* **2020**, *258*, 120775. [CrossRef]
20. Green Service. Available online: http://Www.Greenservice.Lt/En/Bioreactor-Bio-10 (accessed on 5 November 2018).
21. Bio-10 Manual. Available online: http://Www.Greenservice.Lt/Images/Documents/Bio-10-Manual.Pdf (accessed on 5 November 2018).
22. Yang, X.; Liu, E.; Zhu, X.; Wang, H.; Dong, W.; Liu, H.; Liu, X. Impact of composting methods on nitrogen retention and losses during dairy manure composting. *Int. J. Environ. Res. Public Health* **2019**, *16*, 3324. [CrossRef]
23. Iqbal, H.; Garcia-Perez, M.; Flury, M. Effect of biochar on leaching of organic carbon, nitrogen, and phosphorus from compost in bioretention systems. *Sci. Total. Environ.* **2015**, *521–522*, 37–45. [CrossRef]
24. Sommer, S. Effect of composting on nutrient loss and nitrogen availability of cattle deep litter. *Eur. J. Agron.* **2001**, *14*, 123–133. [CrossRef]
25. Boldrin, A.; Andersen, J.K.; Møller, J.; Christensen, T.H.; Favoino, E. Composting and compost utilization: Accounting of greenhouse gases and global warming contributions. *Waste Manag. Res.* **2009**, *27*, 800–812. [CrossRef]
26. Ulén, B. Leaching of plant nutrients and heavy metals during the composting of household wastes and chemical characterization of the final product. *Acta Agric. Scand. Sect. B-Plant Soil Sci.* **1997**, *47*, 142–148. [CrossRef]
27. Goedkoop, M.; Heijungs, R.; Huijbregts, M.; Schryver, A.; Struijs, J.; Zelm, R. Recipe 2008: A Life Cycle Impact Assessment Method Which Comprises Harmonised Category Indicators at the Midpoint and the Endpoint Level; 2008. Available online: https://www.leidenuniv.nl/cml/ssp/publications/recipe_characterisation.pdf (accessed on 8 December 2021).
28. Huijbregts, M.A.J.; Steinmann, Z.J.N.; Elshout, P.M.F.; Stam, G.; Verones, F.; Vieira, M.; Zijp, M.; Hollander, A.; Van Zelm, R. Recipe2016: A harmonised life cycle impact assessment method at midpoint and endpoint level. *Int. J. Life Cycle Assess.* **2017**, *22*, 138–147. [CrossRef]
29. Ecoinvent Database V3.6. Swiss Centre for Life Cycle Inventories. 2019. Available online: https://ecoinvent.org/the-ecoinvent-database/ (accessed on 8 December 2021).
30. Présustainabili Ty. Simapro-Lca Software For Fact-Based Sustainability. Available online: Https://Simapro.Com/ (accessed on 6 March 2019).
31. Kumar, S.; Karan, R.; Kapoor, S.; Singh, S.P.; Khare, S.K. Screening and isolation of halophilic bacteria producing industrially important enzymes. *Braz. J. Microbiol.* **2012**, *43*, 1595–1603. [CrossRef]

32. Gao, M.; Li, B.; Yu, A.; Liang, F.; Yang, L.; Sun, Y. The effect of aeration rate on forced-aeration composting of chicken manure and sawdust. *Bioresour. Technol.* **2010**, *101*, 1899–1903. [CrossRef]
33. Dde Guardia, A.; Mallard, P.; Teglia, C.; Marin, A.; Le Pape, C.; Launay, M.; Benoist, J.; Petiot, C. Comparison of five organic wastes regarding their behaviour during composting: Part 2, nitrogen dynamic. *Waste Manag.* **2010**, *30*, 415–425. [CrossRef]
34. Setia, R.; Marschner, P.; Baldock, J.; Chittleborough, D. Is Co2 evolution in saline soils affected by an osmotic effect and calcium carbonate? *Biol. Fertil. Soils* **2010**, *46*, 781–792. [CrossRef]
35. Haroun, M.; Idris, A.; Omer, S. Composting of Tannery Sludge With Chicken Manure and Rice Bran. *Sci. Technol.* **2004**, *11*, 300–307.
36. Turan, N.G. The effects of natural zeolite on salinity level of poultry litter compost. *Bioresour. Technol.* **2008**, *99*, 2097–2101. [CrossRef] [PubMed]
37. Straathof, A.L.; Comans, R.N. Input materials and processing conditions control compost dissolved organic carbon quality. *Bioresour. Technol.* **2015**, *179*, 619–623. [CrossRef]
38. Gul, S.; Whalen, J.K. Biochemical cycling of nitrogen and phosphorus in biochar-amended soils. *Soil Biol. Biochem.* **2016**, *103*, 1–15. [CrossRef]
39. Li, J.; Shi, W.; Jiang, C.; Bai, L.; Wang, T.; Yu, J.; Ruan, W. Evaluation of potassium as promoter on anaerobic digestion of saline organic wastewater. *Bioresour. Technol.* **2018**, *266*, 68–74. [CrossRef]
40. Pandey, P.K.; Cao, W.; Biswas, S.; Vaddella, V. A new closed loop heating system for composting of green and food wastes. *J. Clean. Prod.* **2016**, *133*, 1252–1259. [CrossRef]
41. Mccauley, A.; Jones, C.; Jacobsen, J. Soil pH and organic matter. *Nutr. Manag. Modul.* **2017**, *8*, 1–16.
42. Valdovska, A.; Vasermane, S. Bioperspektīva mastīta profilaksei govju ganāmpulkā. *Agrotops* **2019**, *11*, 60–62.
43. Rigobelo, E.C.; Cardozo, M.V.; De Avila, F.A.; Blackall, P.J. An evaluation of the use of probiotics and manure composting as strategies to reduce levels of Shiga toxin-producing Escherichia coli in sheep. *Afr. J. Microbiol. Res.* **2016**, *10*, 1011–1017.

processes

MDPI

Article

# The Use of Phosphate Washing Sludge to Recover by Composting the Leachate from the Controlled Landfill

**Meriem Mobaligh [1], Abdelilah Meddich [2], Boujamaa Imziln [3] and Khalid Fares [1,*]**

1 Laboratory of Pharmacology, Neurobiology, Anthropobiology and Environment, Faculty of Sciences Semlalia, Cadi Ayyad University, Marrakech 40000, Morocco; meriem.mobaligh.15@gmail.com

2 Laboratory of Agro-Foods, Biotechnologies and Valorisation of Plants Bioresources, Department of Biology, Faculty of Sciences Semlalia, Cadi Ayyad University, Marrakech 40000, Morocco; meddichabdelilah@yahoo.fr

3 Laboratory of Microbial Biotechnologies, Agroscience and Environment, Department of Biology, Faculty of Sciences Semlalia, Cadi Ayyad University, Marrakech 40000, Morocco; imziln@uca.ac.ma

* Correspondence: fares@uca.ac.ma; Tel.: +212-661-232-285

**Abstract:** The percolation of rainwater and runoff water through household waste in the dumpsite generally leads to an overabundance of leachate in Moroccan landfills, which is a source of soil, surface water and groundwater contamination. In order to ecologically solve the problem posed by the leachate in the dump site, to safeguard the environment and to contribute to sustainable development, we have carried out this study which aims to study the possibility of composting leachate with green waste and phosphate washing sludge. Various combinations with five substrates (leachate, green waste, sugar lime sludge, phosphate washing sludge and olive mill wastewater) in different proportions were used to build five windrows. A 24 h contact between the phosphate sludge or sugar lime sludge and the leachate took place prior to the addition of the green waste for the construction of the different windrows. This contact time ensured the absorption of a significant portion of the leachate and the disappearance of bad odor. A significant reduction was obtained with streptococci and mesophilic flora after 24 h of contact. The monitoring of the physicochemical parameters throughout the composting process showed that the temperature of the different windrows followed a good pace presenting all composting phases. Moisture, pH, C/N ratio and the percentage of degradation of the organic matter conformed to the quality standards of the compost. The combinations of the alkaline treatment and the composting process allowed a significant hygienization of the leachate. The results of the humification parameters and the E4/E6 ratio suggest that the composts obtained with phosphate sludge were the most stable and mature and can be used in the agricultural field or green space.

**Keywords:** olive mill wastewater; leachate; compost; phosphate sludge; physicochemical and microbiological analyzes; sugar lime sludge

**Citation:** Mobaligh, M.; Meddich, A.; Imziln, B.; Fares, K. The Use of Phosphate Washing Sludge to Recover by Composting the Leachate from the Controlled Landfill. *Processes* **2021**, *9*, 1735. https://doi.org/10.3390/pr9101735

Academic Editor: Antoni Sánchez

Received: 19 August 2021
Accepted: 26 September 2021
Published: 28 September 2021



## 1. Introduction

In developing countries, solid waste management remains crucial because of demographic change, forced urbanization, and the improvement of living standards in each country [1]. In Morocco, the daily quantity of household waste generated is estimated at 18,000 tons, and this amount is increasing, with direct and indirect negative effects related to the nature and quantity of waste, its disposal, and treatment. This situation has prompted Morocco to opt for the technique of landfilling as a means of managing these huge quantities of waste characterized by their high organic matter content (65%) and high humidity (85%). However, the storage of these wastes at landfill sites produces harmful effluents (leachate), resulting from the combined action of rainwater and natural fermentation, known for their high loads of organic and mineral substances and thus pathogenic microorganisms, which creates a major environmental problem given the emissions of bad odors, soil and groundwater pollution [2].

Today in Morocco, in addition to the production of household waste, there are three other serious environmental problems such as that of the phosphate industry. It represents one of the main drivers of the country's economy. Morocco, with its large share of phosphate reserves (three-quarters of the world's phosphate reserves), is the world's leading exporter of phosphate derivatives, with an international market share of more than 30% [3]. This mining industry faces many major environmental challenges, resulting from the huge quantities (28.1 million tons per year) of phosphates washing sludge that Cherifian Phosphates Office (OCP) generates [4]. All these by-products are deposited together at the level of basins on the sites of the laundries, where they are stored. Moreover, the sugar industry generates an enormous amount of sugar lime sludge, by-products of the purification of beet juice, which increases each year with the increase in sugar beet and cane production. It is estimated that nearly 270,000 t of sugar lime sludge are produced each year [5]. This huge quantity of sugar lime sludge is always dumped outside the plants in wild lands without any valorization or treatment. Beyond their richness in calcium, phosphoric acid and magnesium assimilated by the plant are also considered perfect to activate agricultural soils and increase their pH [6], hence, the need for their exploitation. The olive industry is another kind of industry that constitutes a real polluting activity due to its production of huge quantities of olive mill wastewater. The discharge of these wastes into the environment without any prior treatment causes negative impacts due to their high organic loads, which are poorly biodegradable and highly toxic to plants, water, and soil microorganisms [7]. So far, these effluents have small economic value in Morocco.

Today, this situation of continuous accumulation of liquid and solid wastes (leachates, phosphate sludge, sugar lime sludge, and olive mill wastewater) encourages researchers to fit perfectly into the challenges of sustainable development by controlling and revalorizing these wastes on a large scale in order to protect the environment and fight against the resulting pollution in nature. From the point of view of the physicochemical characteristics of these substrates, composting seems to be the first effective solution for better ecological management, which will allow developing countries to engage with the stakes of the circular economy, which are at the same time an environmental, economic, and social issue.

In order to improve the performance of the composting process and to optimize the quality of final composts, we opted to carry out a comparison between the effect of using different concentrations of phosphate sludge and sugar lime sludge. Indeed, the physicochemical characteristics of phosphate sludge are almost identical to those of sugar lime sludge with regard to pH, organic matter, and total organic carbon. The main objective of this work was the valorization of phosphate sludge and sugar lime for the treatment by composting of leachates, and the elimination of their bad odors in order to obtain valuable composts that could be used in the agricultural valorization without negative effects on the soil and the crops. This study presents ecological and socio-economic benefits in the context of sustainable development related to the protection of the environment and natural resources through the recycling and valorization of large quantities of leachate.

## 2. Material and Methods

### 2.1. Composting Materials and Process Operation

Five types of waste including leachate (L), green waste (GW), phosphate sludge (PS), sugar lime sludge (SL), and olive mill wastewater (OMW) were used.

These wastes have various origins:

a. Olive mill wastewater is recovered from Bouchane oil mill, Morocco.
b. Green waste (grass) was collected from the garden of the Faculty of Science Semlalia Cadi Ayyad University of Marrakesh, Morocco. This type of green waste was chosen to ensure better absorption of the leachate.
c. The phosphate washing sludge was obtained from the Youssoufia region, Morocco.
d. Leachate was sampled from the ECOMED landfill in Marrakech, Morocco on 2 March 2020.
e. Sugar lime sludge was from the Doukkala sugar industry, Morocco.

The physicochemical properties of the initial substrates used in this work are presented in Table 1.

**Table 1.** Physicochemical properties of the initial substrate.

| Characteristics | L | GW | OMW | PS | SL |
|---|---|---|---|---|---|
| **pH** | 8.0 ± 0.04 | 7.9 ± 0.03 | 5.1 ± 0.01 | 7.9 ± 0.1 | 8.1 ± 0.05 |
| **TKN (% DM)** | 0.4 ± 0.1 | 2.8 ± 0.1 | 0.4 ± 0.1 | 0.04 ± 0.1 | 0.3 ± 0.1 |
| **Organic matter (% DM)** | nd | 81.2 ± 0.6 | nd | 10.7 ± 2.1 | 12.7 ± 0.2 |
| **TOC (% DM)** | nd | 45.1 ± 0.3 | nd | 5.9 ± 1.2 | 7.1 ± 0.1 |
| **Humidity (%)** | 95.7 ± 0.1 | 12.1 ± 0.6 | 90.5 ± 0.1 | 2.4 ± 0.1 | 4.7 ± 0.06 |
| **$BOD_5$ (mg $O_2$ /L)** | 1400 ± 0.0 | nd | nd | Nd | nd |
| **COD (mg $O_2$ /L)** | 25,750 ± 403.7 | nd | 190,550 ± 14,131.3 | Nd | nd |
| **$BOD_5$/COD** | 0.05 | nd | nd | Nd | nd |
| **Ni** | 0.07 ± 0.01 | 0.9 ± 0.3 | 0.01 ± 0.0 | 27.7 ± 1.1 | 0.8 ± 0.2 |
| **Cu** | 0.0 | 12.6 ± 0.1 | 0.0 | 32.4 ± 0.6 | 10.0 ± 0.2 |
| **Pb** | 0.01 ± 0.0 | 1.4 ± 0.1 | 0.01 ± 0.0 | 1.1 ± 0.01 | 0.7 ± 0.02 |
| **Zn** | 0.04 ± 0.0 | 55.8 ± 0.7 | 0.3 ± 0.01 | 302.2 ± 6.2 | 23.0 ± 0.6 |
| **Cr** | 0.07 ± 0.0 | 0.7 ± 0.04 | 0.0 | 51.4 ± 1.0 | 0.9 ± 0.02 |
| **As** | 0.5 ± 0.1 | 0.4 ± 0.1 | 0.0 | 19.6 ± 0.4 | 1.9 ± 0.02 |

L (leachate); GW (green waste); OMW (olive mill wastewater); PS (phosphate sludge); SL (sugar lime); TKN (total Kjeldahl nitrogen); TOC (total organic carbon); $BOD_5$ (biochemical oxygen demand); COD (chemical oxygen demand); Ni ( nickel); Cu (copper); Pb (lead); Zn (zinc); Cr (chromium); As (arsenic).

One day before composting is started (2 March 2020), we prepared five barrels with 80 L leachate in each barrel. This proportion was chosen in order to recover the largest possible quantity of final compost. A barrel control without any treatment was maintained and the other barrels were subjected to different treatments (Table 2) in order to test the effect of the alkaline sludge on the removal of fecal contamination indicators.

**Table 2.** Treatments carried out for each barrel.

| Barrels | Treatments | % of Sludge (*w*/*v*) |
|---|---|---|
| Barrel 1 | L | 0 |
| Barrel 2 | L + SL | 20 |
| Barrel 3 | L + PS | 20 |
| Barrel 4 | L + PS | 50 |
| Barrel 5 | L + PSW | 50 |

L (leachate); L + SL (leachate + sugar lime); L + PS (leachate + phosphate sludge); L + PSW (leachate + phosphate sludge + olive mill wastewater).

Subsequently, the barrels were carefully mixed and incubated 24 h at ambient temperature. This contact time is necessary for the reaction of chemical substances in the phosphate sludge and sugar lime sludge with the leachate. To maintain homogeneity of the mixture, the barrels were mixed five times during the contact time. After 24 h of contact between leachate/phosphate sludge and leachate/sugar lime sludge, we collected about 500 mL of sample from each treatment for physicochemical and microbiological analyses. Then, a sufficient quantity of green waste was added to each barrel in order to absorb the maximum of leachate and obtain a final C/N ratio between 25 and 28. The mixtures from each barrel were carefully mixed by turning to ensure good homogenization and the implementation of the windrows (1 m length, 0.7 m wide, and 0.5 m high) was performed on plastic sheeting to prevent leachate losses. The quantities (in kg) of the different wastes according to the windrows are shown in Table 3.

**Table 3.** Composition of the five windrows in kg.

| Windrows | GW (kg) | PS (kg) | L (L) | SL (kg) |
|---|---|---|---|---|
| W1 | 33 | 0 | 80 | 0 |
| W2 | 33 | 0 | 80 | 16 |
| W3 | 33 | 16 | 80 | 0 |
| W4 | 33 | 40 | 80 | 0 |
| W5 | 33 | 40 | 80 | 0 |

GW (green waste); PS (phosphate sludge); L (leachate); SL (sugar lime).

### 2.2. Monitoring Composting

We conducted regular manual turning twice a week to ensure aeration of the windrows, as the aeration rate is considered one of the key factors affecting the composting process and the final quality of the compost; the windrows were watered whenever the moisture content dropped below 50%. It should be noted that windrow 5 was watered on day 7 and day 21 by olive mill wastewater (7 L in total).

The temperature was measured every day during the first week and once every 3 days for the rest. The value given corresponds to the mean of five measurements taken at different locations and varying depths of each windrow. At predetermined periods (start of composting, 7, 14, 21, 35, 49, 63, 77, 91, and 112 days), sampling was carried out uniformly throughout each windrow, at the surface and at different depths in order to obtain a homogeneous sample for physicochemical and microbiological analyses.

### 2.3. Physicochemical Analysis

The five windrows were characterized according to the main conventional physicochemical parameters monitored during composting. The samples taken were subjected to various analyses, including the moisture content, which was determined on a 100 g sample after drying in an oven at 105 °C for 24 h. The pH was measured in a 1:10 ($w/v$) aqueous extract. The organic matter (OM) content was obtained by calcination at 650 °C for 6 h. The total organic carbon (TOC) was calculated according to Equation (1) [8]:

$$\text{COT } (\%) = \frac{OM\ (\%)}{1.8} \tag{1}$$

Total Kjeldahl nitrogen and assimilable phosphorus were determined according to the Kjeldahl method and the Olsen method, respectively. The heavy metals were determined by ICP-MS. All analyses were carried out in three replicates.

### 2.4. Determination of the Bacterial Population

Microbiological analyses were carried out on the leachate before any addition of sludge, on the leachate after 24 h contact with the sludge, and on the initial and final composting stages.

The samples collected were analyzed for microbial load indicators of fecal contamination and some pathogenic microorganisms. Fecal coliform and total coliform were determined on tergitol 7 and TTC agar. Bile esculin-azid agar was used for the enumeration of fecal streptococci. *Salmonella* spp. was identified using SS agar and confirmed on a triple sugar iron medium. Enumeration of the total mesophilic flora was performed on nutritious agar and diagnosis of *Pseudomonas aeruginosa* was performed using cetrimide agar as a selective medium.

### 2.5. Humification Process

In order to prove the stability and maturity of the five final composts, we calculated the humification indices (humification index (HI), humification rate (HR), and degree of humification (HD)), which are based on the quantification of the humic fraction compared to the fulvic fraction. For this purpose, we carried out the extraction of the total humic substances from the compost according to the method of [9]. The separation of humic (HA) and fulvic (FA) acids was carried out by acidification with a sulphuric acid solution (6 N). The humidification rate, humidification degree, and humidification index are calculated according to Equations (2) [10], (3) [11], and (4) [12] respectively:

$$\text{Humification rate (HR\%)} = (\text{Carbon of total humic matter}/\text{Total organic carbon}) \times 100 \quad (2)$$

$$\text{Humification degree (HD\%)} = (\text{Carbon of total humic matter}/\text{Carbon of Humin}) \times 100 \quad (3)$$

$$\text{Humification index (HI)} = \text{Carbon of humic acid}/\text{Carbon of fulvic acid ratio} \quad (4)$$

### 2.6. Statistical Analysis

All the results are expressed as the mean and standard deviation. The physicochemical and microbiological results were analyzed by the ANOVA using XLStat Premium, version 2013.

## 3. Results and Discussion

### 3.1. Moisture Content

Generally, the moisture content is necessary for the bacterial activity and the degradation of organic matter during composting [13]. According to [14], the optimum moisture content is between 40% and 60%. The evolution of the moisture content of the different windrows during the composting process is shown in Figure 1. The results showed that windrows W1, W2, W3, W4, and W5 started with a moisture content of 73.2%, 67.5%, 64.2%, 58.2%, and 58.4%, respectively. The highest moisture content was recorded for W1, which does not contain any sludge, followed by W2 (20% of lime sludge). In fact, the results obtained in this study showed that phosphate sludge supplementation to the leachate decreased the moisture content as the amount added increased compared to the control. Thus, the phosphate sludge ensures better absorption of the leachate. The moisture content decreases in all windrows progressively to reach at the end of composting 53.3%, 45.5%, 42.2%, 40.7%, and 40.5%, respectively for W1, W2, W3, W4, and W5. This decrease could be attributed to the evaporation due to the increase in temperature generated by the microbial activity during composting [15]. Moreover, it is the consequence of the composting conditions, which we have proceeded to ensure the smooth running of the process, stirring, and aeration that leads to water loss in the form of steam. It should be noted that the difference between the windrow control and the windrow containing sugar lime sludge is significant ($p < 0.05$), but the difference between the windrow control and the windrows containing phosphate sludge is highly significant ($p < 0.01$).

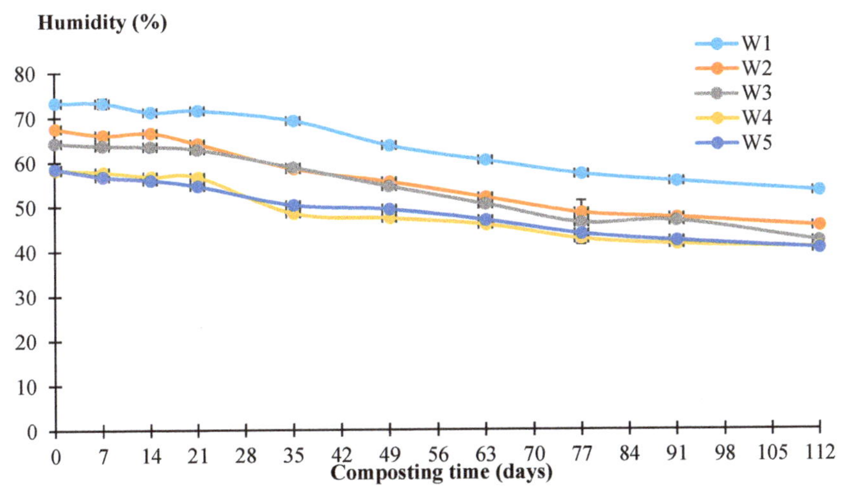


**Figure 1.** Variation of the humidity in the windrows during composting.

### 3.2. Temperature Evolution

According to [16], the success of composting depends on the temperature reached during the process. Indeed, the ability of the composting process to hygienize the compost from pathogens is associated with temperature [17]. The results in Figure 2 showed that the temperature curves for the five windrows had classic composting patterns, but no significant difference was detected between the control and the four different treatments. The thermophilic phase was established on the first day of composting and lasted almost 10 days, reaching maximum values of 54 °C for W1 and W3, followed by 53 °C for W4 and 51 °C for W2 and W5. This rapid increase observed during the initial phase is due to the degradation of simple compounds by microorganisms producing heat as a by-product [18,19]. After this phase, the temperature of the five windrows gradually decreased which may be in relation to the cold days as proved by the values of the ambient temperature. Then it increased again on the 63rd, 77th, and 84th days of composting. From 91 days, the temperature of the five windrows dropped to ambient temperature values, reflecting the maturity and stability of the composts. This indicates that simply metabolizable organic compounds have been decomposed and only compounds resistant to metabolization persist [20].

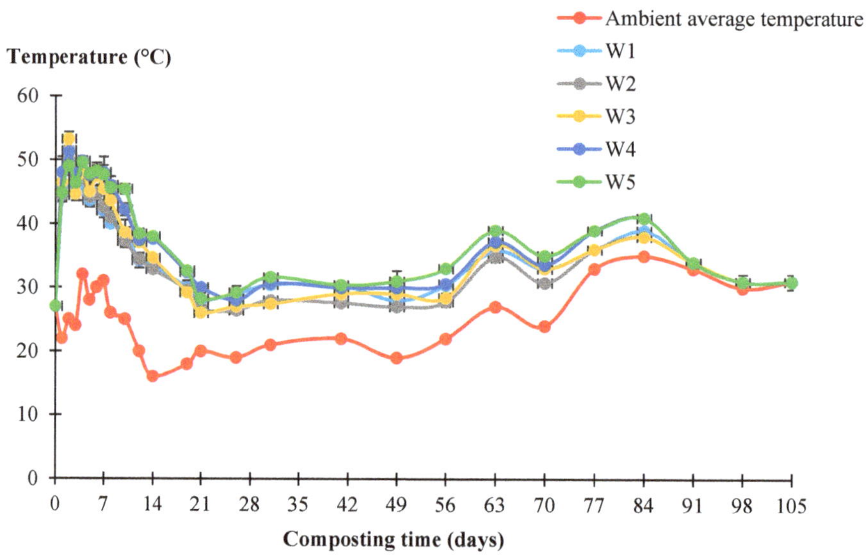


**Figure 2.** Temperature variation during the composting process.

According to [21], this last phase of composting (maturation) is characterized by the humification process, which consists of the polymerization of organic compounds into more stable compounds (humus).

### 3.3. Evolution of the pH

The evolution of pH during composting is shown in Figure 3. The change in pH of windrow 2 compared to the control was not significant, but the change in pH of windrows 3, 4, and 5 compared to the control was highly significant ($p < 0.01$). From the first week of composting, there was a slight decrease in the pH of the five windrows due to the accumulation of organic acids and the dissolution of $CO_2$ during the degradation of simple molecules [22]. The increase in pH for the five windrows in the following week could be explained by the ammonia production associated with the degradation of organic matter [23]. From day 21 to day 35, the pH of the five windrows decreases probably in relation to the volatilization of $NH_3^+$. Afterwards, a slight increase in pH was observed in the five windrows due to the disappearance of easily degradable organic matter [24], and then the pH of the five windrows stabilized at relatively basic values: 8.6, 8.8, 8.8, 8.9, and 8.9 for W1, W2, W3, W4, and W5, respectively. Recommended pH values for mature compost are normally between 7 and 9 [25]. Throughout the composting process, and even with the same concentration of phosphate sludge, windrow 5 showed a lower pH than windrow 4, which could be logically associated with watering by the olive mill wastewater, which has a pH = 5.1 (Table 1).

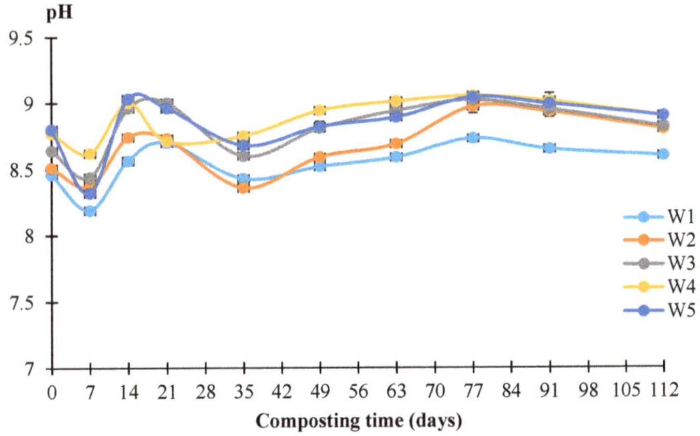


**Figure 3.** Evolution of the pH in the different windrows during the composting process.

*3.4. C/N Ratio Evolution and Organic Matter Degradation*

The C/N ratio is one of the factors used to control the composting process [26]. The C/N ratio decreased gradually and followed an almost similar trend in the five windrows (Figure 4). This is probably due to the loss of carbon in the form of carbon dioxide due to the degradation of organic matter and the increase in total nitrogen content due to organic matter mineralization of the initial substrates by the microorganisms [27]. At the end of composting, the C/N ratio of the final composts was 11.8, 13.0, 13.7, 12.2, and 12.7 for windrows 1, 2, 3, 4, 5, respectively, which proves that the final composts were mature and stable, as they were within the range recommended by [28].

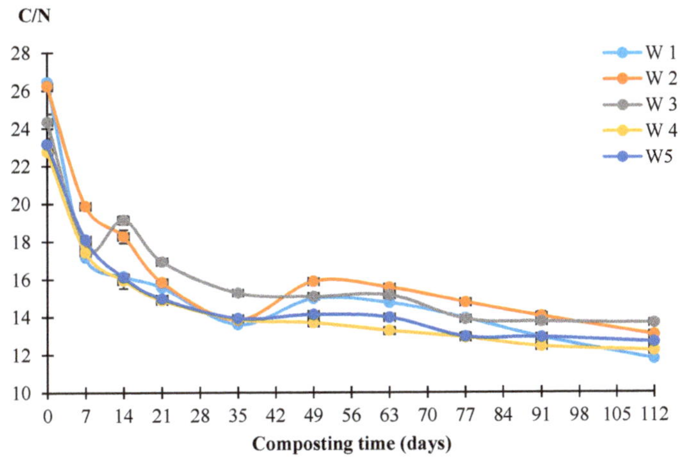


**Figure 4.** Evolution of the C/N ratio during composting.

The organic matter content of the final composts was within the range recommended by [27]. It was in the order of 53.0%, 38.0%, 31.9%, 24.6%, 24.6%, and 25.0% for W1, W2, W3, W4, and W5, respectively. The maximum organic matter degradation (Figure 5) was observed for W4 (34.1%), followed by W3 (33.9%), W5 (32.5%), and W2 (30.1%), while for W1, which did not contain any sludge, the organic matter degradation was only 15.1%. These results showed that the addition of phosphate sludge or sugar lime sludge had a significant impact on the degradation of organic matter. The percentage of organic matter degradation was not similar in the two windrows W4 and W5 that underwent the same treatment (50% PS) due to the fact that windrow 5 was watered by olive mill wastewater on day 7 by 5 L and day 21 by 2 L. This result showed that the 7 L watering by the olive mill wastewater contributed to a slight increase in organic matter in windrow 5 of about 1.9% DM compared to windrow 4.

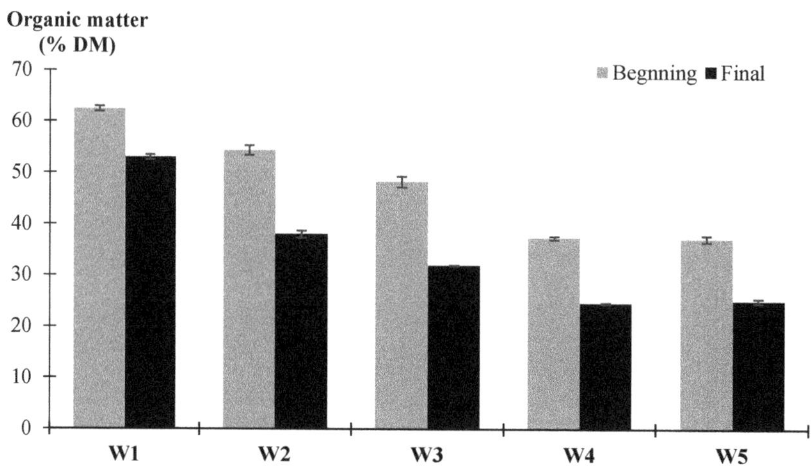


**Figure 5.** Organic matter contents in the different windrows.

### *3.5. Monitoring of Microbiological Parameters*

#### 3.5.1. Microbiological Characteristics of the Raw Leachate

Microbiological analyses of the raw leachate (Table 4) used in this study showed a total absence of *Salmonella* spp. and *Pseudomonas aeruginosa*. The absence of these pathogenic microorganisms could be explained, among other things, by the alkaline nature of the leachate, which is unfavorable for the development of these bacteria [29]. Total and fecal coliforms are frequent in the environment. However, their absence in the leachate sample may be due to the absence of favorable conditions for their development. Indeed, results relating to the evolution of fecal coliforms suggest that these microorganisms are seasonally dependent: a decrease in the bacterial load of fecal coliforms during the cold season has been recorded [30].

**Table 4.** Results of the microbiological analyses on raw leachate from the landfill of Marrakech.

| Microorganismes | CFU/mL |
|---|---|
| Fecal streptococci | 9050 |
| Fecal coliforms | 0 |
| Total coliforms | 0 |
| Total mesophilic flora | 163,500,000 |
| *Pseudomonas aeruginosa* | 0 |
| *Salmonella* spp. | 0 |

An abundance of fecal streptococci was detected in the leachate sample. Indeed, fecal streptococci are very good indicators of fecal contamination and are more resistant to environmental factors than coliforms [31]. A high concentration of mesophilic flora was also observed. In general, the total mesophilic flora provides information on the indigenous microflora brought by pollution; it is used as an indicator of overall pollution [30].

#### 3.5.2. The impact of Adding Phosphate Sludge on the Content of Microorganisms in the Leachate

Before starting the composting process, we proceeded to carry out a direct contact of 24 h between the sludge and the leachate, in order to evaluate the effect of the different treatments on the bacterial load of the raw leachate.

The addition of phosphate sludge and sugar lime sludge allowed a significant reduction in fecal streptococci and total mesophilic flora (Figures 6 and 7); almost similar trends were observed for the two treatments phosphate sludge 20% and sugar lime sludge 20%. With 20% sugar lime sludge, a 53.7% reduction in fecal streptococci was observed, while with 20% of the phosphate sludge, a 51.9% destruction was observed after 24 h of contact. When the phosphate sludge addition was increased to 50%, 56.4% to 57.8% destruction of fecal streptococci was observed.

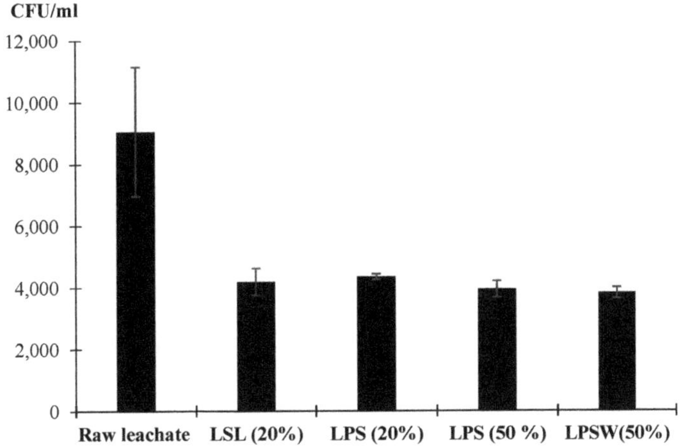


**Figure 6.** Evolution of the fecal streptococci in the different treatments after 24 h of contact.

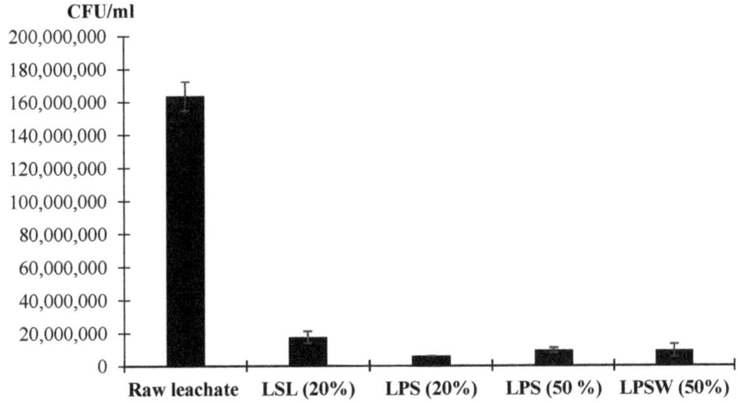


**Figure 7.** Evolution of the total mesophilic flora in the different treatments after 24 h of contact.

The concentration of the total mesophilic aerobic flora was considerably destroyed after 24 h of contact for all four treatments: 89.3%, 96.3%, 98.6%, and 98.4% for W-SL (20%), L-PS (20%), L-PS (50%), and L-PSW (50%), respectively. The percentage of reduction in total mesophilic flora was higher than that of fecal streptococci, indicating its sensitivity to lime.

The treatments have successfully reduced the pathogenic agents. These results confirm those found by [5], which demonstrated the effectiveness of alkaline treatments in pathogen reduction. Although pathogen reduction was satisfactory, a complete composting cycle was necessary for better hygienization of the final products.

#### 3.5.3. The Effect of Composting on the Hygienization

To evolve the effect of composting as a hygienization process, we conducted counts of the microbial community considered at the beginning and the end of the composting process. Although there was a significant drop in the total mesophilic flora due to the addition of phosphate sludge and sugar lime sludge, the addition of green waste was satisfactory in terms of increasing the total mesophilic flora (as well as fecal streptococci) required to start composting Figure 8. At the beginning of composting, the total mesophilic flora was predominant: $1 \times 10^9$ CFU/g for W1, $8.45 \times 10^8$ CFU/g for W2, $7.8 \times 10^8$ CFU/g for W3, $6.5 \times 10^8$ CFU/g for W4, and $5.65 \times 10^8$ CFU/g for W5. Towards the end of composting, this flora remained abundant in the final composts: $1.6 \times 10^8$ CFU/g, $1.32 \times 10^8$ CFU/g, $8.73 \times 10^7$ CFU/g, $1.03 \times 10^8$ CFU/g, and $1.07 \times 10^8$ CFU/g for W1, W2, W3, W4, and W5, respectively. This could be explained by the environmental conditions that favor the reinstallation of the new mesophilic flora, which is essential for obtaining a mature product [32]. Although there was a reduction in fecal streptococci in the control (90%), a high level of fecal streptococci ($2.83 \times 10^3$ CFU/g compost) was still observed even after 112 days of composting. However, treatment with sugar lime sludge resulted in a greater reduction than the control (97.5%). No fecal streptococci were detectable after 112 days in the three composts treated with phosphate sludge. This reduction could be attributed to the alkaline treatment and the hygienization process known in the composting operation. The study of [33] considered the increase in temperature and the increase in pH that leads to the release of ammonia as the main factors involved in the reduction of pathogens from biosolids during composting. We can conclude that the combination of the alkaline treatment and the composting process stabilized the final products to meet those recommended in [28].

### 3.6. *Humification Process during Composting*

According to [34], the monitoring of humification parameters during composting could be a representative index of the evolution of compost maturity and stability, since compost quality is confirmed by the quantity of stable humus formed after the biodegradation of organic matter [35]. The results reported in Table 5 showed that at the end of composting (112 days), all composts containing phosphate sludge had their degree of humification and their humification rate above 70% and 35%, respectively, which are considered standards, indicating a good humification [36]. Similarly, the CAH/CFA ratio was 1.9 for compost 3, 3.5 for compost 4, and 3.0 for compost 5. These values were higher than the 1.6 value estimated by [37] for stabilized organic matter.

In contrast, the E4/E6 ratio for the composts treated with phosphate sludge and sugar lime sludge is less than five, indicating that the composts are mature and its humus particles are complex [38]. The E4/E6 ratio of the control is higher than five, which proves that the degradation of organic matter in the control (without sludge) was less advanced and that the humic substances formed are not stable. According to [39], an E4/E6 ratio higher than five shows the presence of fulvic and the formation of small molecules.

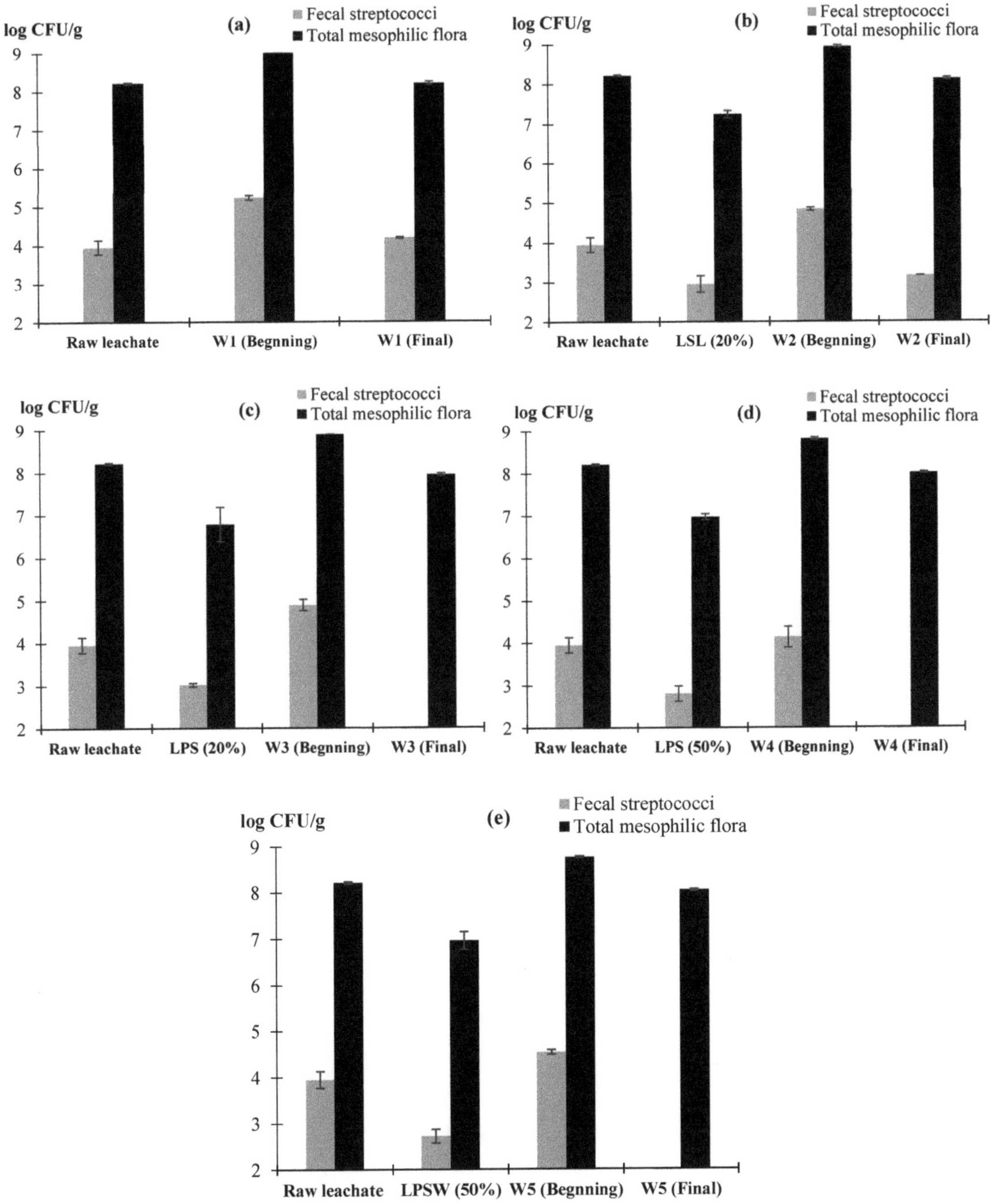


**Figure 8.** Results of microbiological analyses on the windrow 1 (**a**), windrow 2 (**b**), windrow 3 (**c**), windrow 4 (**d**), and windrow 5 (**e**) at the beginning and after 112 days of composting.

**Table 5.** Humification indices of the five final composts.

| Composts | C1 | C2 | C3 | C4 | C5 |
|---|---|---|---|---|---|
| Humification index (IH) | 0.9 ± 0.1 | 1.3 ± 0.2 | 1.9 ± 0.2 | 3.5 ± 0.6 | 3.0 ± 0.2 |
| Humification rate (HR%) | 27.0 ± 0.7 | 41.7 ± 0.6 | 53.9 ± 0.6 | 78.1 ± 2.4 | 76.6 ± 0.2 |
| Humification degree (HD%) | 43.6 ± 0.0 | 59.2 ± 5.5 | 78.4 ± 2.4 | 95.7 ± 0.3 | 97.8 ± 2.7 |
| E4/E6 | 11.8 ± 0.6 | 3.5 ± 0.3 | 2.7 ± 0.1 | 3.5 ± 0.2 | 3.7 ± 0.1 |

According to these results, the addition of phosphate sludge has a significant effect on the humification index, the humification rate, the degree of humification, and the E4/E6 ratio compared to the control and even compared to the compost treated with sugar lime sludge. Therefore, the addition of phosphate sludge allowed obtaining a more stable and mature compost compatible with the use in agricultural fields or green spaces.

### 3.7. *Effect of the Composting Process on Available Phosphorus Content*

As shown in Figure 9, a significant increase in available phosphorus content from the beginning to the end of the composting process was recorded in the five composts. This increase was in the order of 34.1%, 34.7%, 31.4%, 53.5%, and 40.7% for C1, C2, C3, C4, and C5, respectively. These results could be explained by the activity of phosphorus solubilizing microorganisms during the composting process to release more assimilable phosphorus than it was in the initial substrate. The difference in average available phosphorus content at the beginning and final stages of the process could also be attributed to the initial organic matter composition of each windrow (Figure 5), which directly influences the available phosphorus content in each compost. According to [40], the form and rate of phosphorus change during the composting process is influenced by the type of waste used to prepare the compost and the degradation of organic matter by microorganisms.

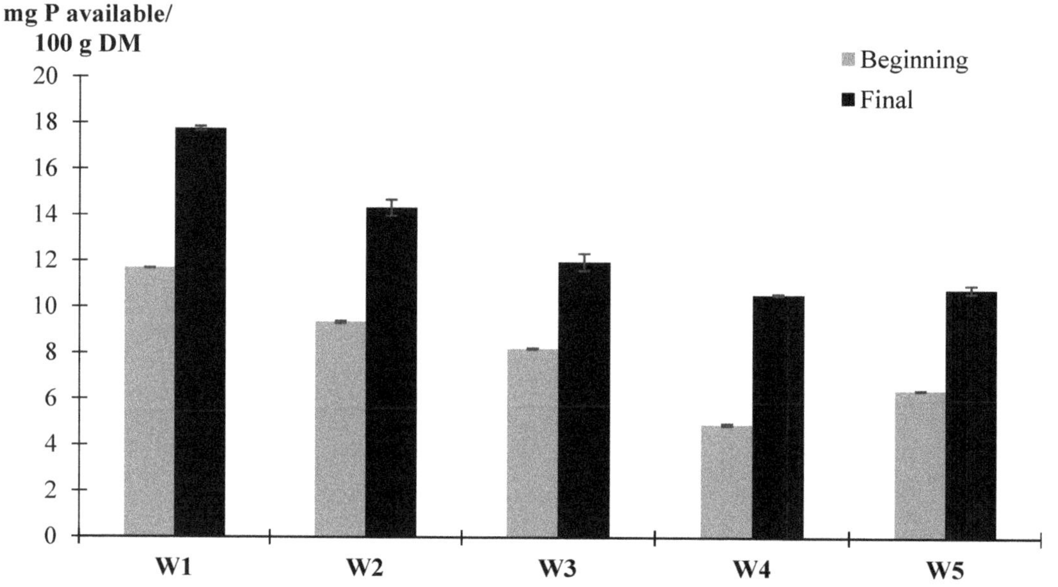


**Figure 9.** Available phosphorus content in the five windrows at the initial and final phases of the composting process.

These results require a complete and detailed speciation of phosphorus and its dynamics during composting to understand its evolution and bioavailability.

### 3.8. Visual Quality of the Final Composts

The results of Table 6 show the effect of the addition of phosphate sludge and sugar lime sludge on the final quality of the composts. For the control (C1), which was not treated, the color was the darkest compared to the other composts. The two composts obtained with 20% of sugar lime sludge (C2) and phosphate sludge (C3) had a similar color (brown). However, by increasing the concentration of phosphate sludge, the final color of the compost 4 can be less brown. Watering windrow 5 with the olive mill wastewater modified the final color of the compost 5. Compared to the control (C1), the treatment of 50% of phosphate sludge ensured a finer texture similar to that of soil characterizing a good degradation of the material during the composting cycle, followed by the two other treatments of 20% of phosphate sludge and sugar lime sludge. The unpleasant odor of the leachate has completely disappeared. In addition, there was a complete absence of impurities (plastic, glass, etc.) in the five composts as stipulated by [28].

**Table 6.** Macroscopic observations on the final composts.

| | **Compost 1 (C1)** | **Compost 2 (C2)** | **Compost 3 (C3)** | **Compost 4 (C4)** | **Compost 5 (C5)** |
|---|---|---|---|---|---|
| Texture | Fine to coarse | Fine to coarse | Fine to coarse | Fine to coarse | to mostly fine |
| Color | Dark brown | Brown | Brown | Light brown | Brown |
| Bad odor | Absent | Absent | Absent | Absent | Absent |

### 3.9. Heavy Metals Concentration

The concentration of heavy metals is considered among the main factors that affect the quality of compost and limit its use and marketing [41]. During the composting the concentration of heavy metals may decrease or increase [42], because the water evaporation which occurs during composting and the solubilization of heavy metals [43] influenced by the physicochemical changes (pH, CEC, $NH_4^+$, $NO_3^-$).

Six heavy metals: nickel, chromium, copper, zinc, lead, and arsenic were analyzed at the beginning and at the end of the process. The results (Table 7) revealed that the heavy metal contents of the windrows varied from one treatment to another. The increase observed in Cu, Zn, Ni, Cr, and As contents towards the end of the process could be due to weight loss due to the degradation of organic matter and loss through respiration during the composting processes [43,44]. A similar result for the increase in Zn, Cr, and Cu has been shown by [41]. On the contrary, Pb concentration decreased towards the end of composting, with the exception of windrows 2 and 3 where an increase was observed. This result is in agreement with that of [45] who showed that when $SO_4^{2-}$ and $PO_4^{3-}$ increase the lead content decreases and also the results obtained by [46] and [41] who found that the lead content increases during composting. The concentrations of heavy metals analyzed in the five final windrows (produced composts) were very low compared to the range concentrations defined by the [28]. Therefore, the composts produced do not contain heavy metals in proportions probable to present a risk for soil, plant, or groundwater contamination.

**Table 7.** Heavy metals concentration in windrows compared to the French Norm (NF U44-051).

| Hevay Metals | Windrows | Begnning | Final | NF U44-051 Range |
|---|---|---|---|---|
| As | W1 | 3.8 ± 0.02 | 8.9 ± 0.02 | 18 mg/kg DM |
| | W2 | 3.4 ± 0.04 | 5.3 ± 0.01 | |
| | W3 | 6.5 ± 0.04 | 7.7 ± 0.1 | |
| | W4 | 5.2 ± 0.1 | 7.3 ± 0.04 | |
| | W5 | 4.0 ± 0.03 | 6.6 ± 0.3 | |
| Cr | W1 | 1.5 ± 0.2 | 5.4 ± 1.9 | 120 mg/kg DM |
| | W2 | 1.6 ± 0.3 | 2.2 ± 0.1 | |
| | W3 | 13.1 ± 1.2 | 10.7 ± 1.8 | |
| | W4 | 13.8 ± 2.0 | 14.9 ± 1.3 | |
| | W5 | 10.4 ± 0.7 | 13.8 ± 0.8 | |
| Zn | W1 | 48.6 ± 0.6 | 95.9 ± 5.5 | 600 mg/kg DM |
| | W2 | 39.8 ± 1.0 | 50.5 ± 1.6 | |
| | W3 | 103.5 ± 1.0 | 125.6 ± 1.6 | |
| | W4 | 98.3 ± 3.4 | 132.7 ± 2.4 | |
| | W5 | 79.8 ± 1.2 | 121.6 ± 10.5 | |
| Cu | W1 | 12.7 ± 0.1 | 26.2 ± 1.0 | 300 mg/kg DM |
| | W2 | 8.5 ± 0.2 | 12.9 ± 0.1 | |
| | W3 | 18.5 ± 0.2 | 17.7 ± 0.1 | |
| | W4 | 12.9 ± 0.6 | 16.8 ± 0.3 | |
| | W5 | 10.2 ± 0.1 | 16.4 ± 0.6 | |
| pb | W1 | 2.3 ± 0.2 | 1.4 ± 0.1 | 180 mg/kg DM |
| | W2 | 0.5 ± 0.04 | 0.9 ± 0.04 | |
| | W3 | 0.4 ± 0.05 | 0.7 ± 0.03 | |
| | W4 | 1.9 ± 0.1 | 0.6 ± 0.02 | |
| | W5 | 2.4 ± 0.1 | 0.8 ± 0.04 | |
| Ni | W1 | 1.8 ± 0.1 | 6.3 ± 0.8 | 60 mg/kg DM |
| | W2 | 2.2 ± 0.3 | 2.9 ± 0.2 | |
| | W3 | 11.0 ± 0.5 | 7.3 ± 0.2 | |
| | W4 | 10.3 ± 1.5 | 8.9 ± 0.3 | |
| | W5 | 7.1 ± 0.2 | 9.2 ± 1.3 | |

## 4. Conclusions

Incubation of leachate in the presence of phosphate washing sludge and sugar lime sludge for 24 h allowed reducing successfully the bacterial load brought by the raw leachate and the removal of bad odors. The addition of the phosphate sludge with both 20% and 50% concentrations and the treatment of the 20% sugar lime sludge optimized significantly the moisture content of the windrows within the recommended range and increased the degradation of organic matter compared to the untreated control windrow. The pH, C/N ratio, microbiological quality, and humification parameters are within the standards of mature and stable composts. Notable results in terms of compost characteristics were recorded with phosphate sludge at 50% concentration compared to the 20% concentration. The treatment with phosphate sludge (20%) had a positive effect on compost parameters, hygienic quality, and humification process compared to the treatment of sugar lime sludge (20%). The use of phosphate washing sludge to valorize leachate from dump site by the production of compost seems to be a sustainable solution, which allows by the same way the valorization of phosphate washing sludge, which remains not valorized in many countries producing phosphates. Following these results, phytotoxicity tests, analyses of radioactivity in the different initial substrates used, as well as the final composts will be carried out to prove the quality of the composts from a toxicity point of view.

**Author Contributions:** M.M. roles/writing—original draft and formal analysis; B.I. and A.M. validation; K.F. supervision and writing—review and editing. All authors have read and agreed to the published version of the manuscript.

**Funding:** This research received no external funding.

**Institutional Review Board Statement:** Not applicable.

**Informed Consent Statement:** Not applicable.

**Data Availability Statement:** Not applicable.

**Conflicts of Interest:** The authors declare no conflict of interest.

## References

1. Charnay, F. Compostage des Déchets Urbains dans les Pays en Développement: Élaboration d'une Démarche Méthodologique Pour une Production Pérenne de Compost. Ph.D. Thesis, Université de Limoges, Limoges, France, 2005. Ecole Doctorale Science—Technique—Santé. p. 5, 277p.
2. Hakkou, R.; Wahbi, M.; Bachnou, A.; El Amari, K.; Hanich, L.; Hibti, M. Impact of Marrakech (Morocco) municipal landfill on water resources. *Bull. Eng. Geol. Environ.* **2001**, *60*, 325–336. [CrossRef]
3. Hakkou, R.; Benzaazoua, M.; Bussière, B. Valorization of Phosphate Waste Rocks and Sludge from the Moroccan Phosphate Mines: Challenges and Perspectives. *Procedia Eng.* **2016**, *138*, 110–118. [CrossRef]
4. Loutou, M.; Hajjaji, M.; Mansori, M.; Favotto, C.; Hakkou, R. Phosphate sludge: Thermal transformation and use as lightweight aggregate material. *J. Environ. Manag.* **2013**, *130*, 354–360. [CrossRef] [PubMed]
5. Saadani Hassani, O.; Saadaoui, N.; Loqman, S.; Rida, S.; Soraa, N.; R'zina, Q.; Fares, K. Impact of carbonatation lime addition on composting of vegetable oil refining sludge. *Sugar Ind.* **2019**, *144*, 139–146. [CrossRef]
6. Asadi, M. *Beet-Sugar Handbook*; John Wiley & Sons: Hoboken, NJ, USA, 2006.
7. Kestioğlua, K.; Yonar, T.; Azbar, N. Feasibility of physicochemical treatment and Advanced Oxidation Processes (AOPs) as a means of pretreatment of olive mill effluent (OME). *Process Biochem.* **2005**, *40*, 2409–2416. [CrossRef]
8. Haug, R.T. *The Practical Handbook of Compost Engineering*; Lewis Publishers: Boca Raton, FL, USA, 1993.
9. Schnitzer, M. Organic matter characterization. In *Methods of Soil Analysis, Part 2. Chemical and Microbiological Properties*, 2nd ed.; Page, A.L., Miller, R.H., Keeney, D.R., Eds.; Agronomy Monograph, N°9; American Society of Agronomy: Madison, WI, USA, 1982; pp. 581–595.
10. Ciavatta, C.; Antisari, L.V.; Sequi, P. A 1st approach to the characterization of the presence of humified materials in organic fertilizers. *Agrochimica* **1988**, *32*, 510–517.
11. Sequi, P.; De Nobili, M.; Leita, L.; Cercignani, G. A new index of humification. *Agrochimica* **1986**, *30*, 175–180.
12. Sugahara, K.; Inoko, A. Composition analysis of humus and characterization of humic acid obtained from city refuse compost. *Soil Sci. Plant Nutr.* **1981**, *27*, 213–224. [CrossRef]
13. Zorpas, A.A.; Loizidou, M. Sawdust and natural zeolite as a bulking agent for improving quality of a composting product from anaerobically stabilized sewage sludge. *Bioresour. Technol.* **2008**, *99*, 7545–7552. [CrossRef] [PubMed]
14. Chowdhury, A.K.M.M.B.; Akratos, C.S.; Vayenas, D.V.; Pavlou, S. Olive mill waste composting: A review. *Int. Biodeterior. Biodegrad.* **2013**, *85*, 108–119. [CrossRef]
15. Khalib, S.N.B.; Zakarya, I.A.; Tengku Izhar, T.N. Composting of Garden Waste using Indigenous Microorganisms (IMO) as Organic Additive. *Int. J. Integr. Eng.* **2018**, *10*, 140–145. [CrossRef]
16. Goyal, S.; Dhull, S.; Kapoor, K. Chemical and biological changes during composting of different organic wastes and assessment of compost maturity. *Bioresour. Technol.* **2005**, *96*, 1584–1591. [CrossRef]
17. Bustamante, M.A.; Paredes, C.; Marhuenda-Egea, F.C.; Pérez-Espinosa, A.; Bernal, M.P.; Moral, R. Co-composting of distillery wastes with animal manures: Carbon and nitrogen transformations in the evaluation of compost stability. *Chemosphere* **2008**, *72*, 551–557. [CrossRef]
18. Nolan, T.; Troy, S.M.; Healy, M.G.; Kwapinski, W.; Leahy, J.J.; Lawlor, P.G. Characterization of compost produced from separated pig manure and a variety of bulking agents at low initial C/N ratios. *Bioresour. Technol.* **2011**, *102*, 7131–7138. [CrossRef]
19. Li, H.; Meng, F.; Duan, W.; Lin, Y.; Zheng, Y. Biodegradation of phenol in saline or hypersaline environments by bacteria: A review. *Ecotoxicol. Environ. Saf.* **2019**, *184*, 109658. [CrossRef]
20. Ait Baddi, G.; Antonio Alburquerque, J.; Gonzálvez, J.; Cegarra, J.; Hafidi, M. Chemical and spectroscopic analyses of organic matter transformations during composting of olive mill wastes. *Int. Biodeterior. Biodegrad.* **2004**, *54*, 39–44. [CrossRef]
21. Mustin, M. *Le Compost, Gestion de la Matière Organique*; François Dubux: Paris, France, 1987.
22. El Fels, L.; Zamama, M.; El Asli, A.; Hafidi, M. Assessment of biotransformation of organic matter during co-composting of sewage sludge-lignocelullosic waste by chemical, FTIR analyses, and phytotoxicity tests. *Int. Biodeterior. Biodegrad.* **2014**, *87*, 128–137. [CrossRef]
23. Vergnoux, A.; Guiliano, M.; Le Dréau, Y.; Kister, J.; Dupuy, N.; Doumenq, P. Monitoring of the evolution of an industrial compost and prediction of some compost properties by NIR spectroscopy. *Sci. Total Environ.* **2009**, *407*, 2390–2403. [CrossRef] [PubMed]

24. McKinley, V.L.; Vestal, J.R. Biokinetic analyses of adaptation and succession: Microbial activity in composting municipal sewage sludge. *Appl. Environ. Microbiol.* **1985**, *47*, 933–941. [CrossRef] [PubMed]
25. Sasaki, N.; Suehara, K.; Kohda, J.; Nakano, Y.; Yano, T. Effects of C/N ratio and pH of raw materials on oil degradation, efficiency in a compost fermentation process. *J. Biosci. Bioeng.* **2003**, *96*, 47–52. [CrossRef]
26. Espinosa-Valdemar, R.M.; Sotelo-Navarro, P.X.; Quecholac-Piña, X.; García-Rivera, M.A.; Beltrán-Villavicencio, M.; Ojeda-Benítez, S.; Vázquez-Morillas, A. Biological recycling of used baby diapers in a small-scale composting system. *Resour. Conserv. Recycl.* **2014**, *87*, 153–157. [CrossRef]
27. Raj, D.; Antil, R.S. Evaluation of maturity and stability parameters of composts prepared from agroindustrial wastes. *Bioresour. Technol.* **2011**, *102*, 2868–2873. [CrossRef]
28. AFNOR; NFU 44-051. Amendements Organiques—Dénomination, Spécifications et Marquage. 2006. Available online: https://www.boutique.afnor.org/fr-fr/norme/nf-u44051/amendements-organiques-denominations-specifications-et-marquage/fa125064/754 (accessed on 26 September 2021).
29. Yosr, S.; Jalel, B.; Sami, S. Combination of air stripping and biological processes for landfill leachate treatment. *Environ. Eng. Res.* **2019**, *25*, 80–87. [CrossRef]
30. Abed, H.; Esamil, A.; Barrahi, M.; Chahboun, N.; Khadmaoui, A.; Ouhssine, M. Analytical evaluation of the microbiological water quality of landfill leachate Kenitra. *Int. J. Innov. Appl. Stud.* **2014**, *7*, 1225–1231.
31. Belle, E.; Genevois, V.; Mudry, J.; Aleya, L. Distribution annuelle de bactéries indicatrices au niveau de la décharge d'ordures ménagères d'Etueffont (France). *Environ. Technol.* **2008**, *29*, 207–216. [CrossRef] [PubMed]
32. Bolta, S.V.; Mihelic, R.; Lobnik, F.; Lestan, D. Microbial community structure during composting with and without mass inocula. *Compost Sci. Util.* **2003**, *11*, 6–15. [CrossRef]
33. Wong, J.W.C.; Selvam, A. Reduction of indicator and pathogenic microorganisms in pig manure through fly ash and lime addition during alkaline stabilization. *J. Hazard. Mater.* **2009**, *169*, 882–889. [CrossRef]
34. López, M.; Elorrieta, M.; Vargas-García, M.; Suárez-Estrella, F.; Moreno, J. The effect of aeration on the biotransformation of lignocellulosic wastes by white-rot fungi. *Bioresour. Technol.* **2002**, *81*, 123–129. [CrossRef]
35. Domeizel, M.; Khalil, A.; Prudent, P. UV spectroscopy: A tool for monitoring humification and for proposing an index of the maturity of compost. *Bioresour. Technol.* **2004**, *94*, 177–184. [CrossRef] [PubMed]
36. Ciavatta, C.; Francioso, O.; Montecchio, D.; Cavani, L.; Grigatti, M. Use of organic wastes of agroindustrial and municipal origin for soil fertilisation: Quality criteria for organic matter. In Proceedings of the First International Encounter, Management of Organic Wastes in Rural Mediterranean Areas, Pamplona, Spain, 22–23 February 2001; pp. 1–17.
37. Jimenez, E.I.; Garcia, V.P. Composting of domestic refuse and sewage sludge. II. Evolution of carbon and some "humification" indexes. *Resour. Conserv. Recycl.* **1992**, *6*, 243–257. [CrossRef]
38. Flaig, W.; Beutelspacher, H.; Rietz, E. Chemical Composition and Physical Properties of Humic Substances. In *Soil Components*; Springer: Berlin/Heidelberg, Germany, 1975; pp. 1–211. [CrossRef]
39. Chen, Y.; Senesi, N.; Schnitzer, M. Information Provided on Humic Substances by E4/E6 Ratios1. *Soil Sci. Soc. Am. J.* **1977**, *41*, 352. [CrossRef]
40. Wei, Y.; Zhao, Y.; Xi, B.; Wei, Z.; Li, X.; Cao, Z. Changes in phosphorus fractions during organic wastes composting from different sources. *Bioresour. Technol.* **2015**, *189*, 349–356. [CrossRef]
41. Tibu, C.Y.; Annang, T.; Solomon, N.; Yirenya-Tawiah, D. Effect of the composting process on physicochemical properties and concentration of heavy metals in market waste with additive materials in the Ga West Municipality, Ghana. *Int. J. Recycl. Organ. Waste Agric.* **2019**, *8*, 393–403. [CrossRef]
42. Lazzari, L.; Sperni, L.; Bertin, P.; Pavoni, B. Correlation between inorganic (PCBs and PAHs) micropollutant concentrations during sewage sludge composting process. *Chemosphere* **2000**, *41*, 427–435. [CrossRef]
43. Amir, S.; Hafidi, M.; Merlina, G.; Revel, J.C. Sequential extraction of heavy metals during composting of sewage sludge. *Chemosphere* **2005**, *59*, 801–810. [CrossRef] [PubMed]
44. Canaruttto, S.; Petruzzelli, G.; Lubrano, L.; Guidi, G.V. How composting affects heavy metal content. *BioCycle* **1991**, *32*, 48–50.
45. Businelli, M.; Gigliotti, G.; Guisquiani, I. Trace element fate in soil profile and corn plant after massive application of urban waste compost: A six-year study. *Agrochimica* **1996**, *40*, 145–152.
46. Pascual, J.A.; Ayuso, M.; Garcia, C.; Hernández, T. Characterization of urban wastes according to fertility and phytotoxicity parameters. *Waste Manag. Res.* **1997**, *15*, 103–112. [CrossRef]

processes

MDPI

Article

# Advanced Characterization of Organic Matter Decaying during Composting of Industrial Waste Using Spectral Methods

Saloua Biyada [1,*], Mohammed Merzouki [1], Hamada Imtara [2,*], Mohamed F. Alajmi [3], Karima Elkarrach [1], Hamza Mechchate [1,4], Raffaele Conte [5] and Mohamed Benlemlih [1]

1 Laboratory of Biotechnology, Environment, Agrifood and Health, Faculty of Sciences Dhar El Mahraz, BP: 1796, Atlas, Sidi Mohamed Ben Abdellah University, Fez 30 000, Morocco; mohammed.merzouki@usmba.ac.ma (M.M.); karima.elkarrach@usmba.ac.ma (K.E.); Hamza.mechchate@usmba.ac.ma (H.M.); Mohamed.benlemlih@usmba.ac.ma (M.B.)
2 Faculty of Arts and Sciences, Arab American University Palestine, Jenin P.O. Box 240, Palestine
3 Department of Pharmacognosy, College of Pharmacy, King Saud University, Riyadh 11451, Saudi Arabia; malajmii@ksu.edu.sa
4 Laboratory of Inorganic Chemistry, Department of Chemistry, P.O. Box 55, FI-00014 Helsinki, Finland
5 Research Institute on Terrestrial Ecosystems (IRET)—CNR, Via Pietro Castellino 111, 80131 Naples, Italy; Raffaele.conte86@tiscali.it
* Correspondence: saloua.biyada@usmba.ac.ma (S.B.); hamada.tarayrah@gmail.com (H.I.)

**Abstract:** To date, compost maturation monitoring is carried out by physical-chemical and microbiological analysis, which could be considered an overweening consumption of time and products. Nowadays, spectroscopy is chosen as a simple tool for monitoring compost maturity. In the present investigation, spectroscopy analysis was performed in the interest of corroborating the compost maturity. This goal was achieved by using the X-ray diffraction, infrared spectroscopy, and scanning electron microscopy. X-ray diffraction analysis showed the presence of the cellulose fraction in compost samples. At the same time, the intensity of pics decreased depending on composting time, thus proving that there was organic matter degradation. Infrared and scanning electron microscopy analysis allow for confirming these results. The correlation between spectroscopies analysis and physical-chemical properties was employed by partial least squares-regression (PLS-R) model. PLS-R model was applied to build a model to predict the compost quality depending on the composting time, the results obtained show that all the parameters analysis are well predicted. The current study proposed that final compost was more stabilized compared with the initial feedstock mixture. Ultimately, spectroscopy techniques used allowed us to confirm the physical-chemical results obtained, and both of them depict maturity and stability of the final compost, thus proving that spectral techniques are more reliable, fast, and promising than physical-chemical analyses.

**Keywords:** compost; textile waste; cellulose; spectroscopy analysis; scanning electron microscopy; PLS-R; maturity

**Citation:** Biyada, S.; Merzouki, M.; Imtara, H.; Alajmi, M.F.; Elkarrach, K.; Mechchate, H.; Conte, R.; Benlemlih, M. Advanced Characterization of Organic Matter Decaying during Composting of Industrial Waste Using Spectral Methods. *Processes* **2021**, *9*, 1364. https://doi.org/10.3390/pr9081364

Academic Editor: Antoni Sánchez

Received: 19 June 2021
Accepted: 28 July 2021
Published: 4 August 2021



## 1. Introduction

Over the past decades, the rising wealth of industry production has yielded a larger amount of waste produced each year. One of the most difficult environmental tasks for scientists is undoubtedly the management and disposal of industrial solid waste [1]. In Morocco, the textile industries are considered one of the industries producing a huge amount of solid waste, and that is disposed of directly into the environment without any treatment [2]. These wastes are a fountainhead of organic matter, which can be absolutely bio-converted into an organic fertilizer, thus contributing to enhance the soil quality [3], and therefore, reduce the risk of erosion if they are treated properly [4].

For many years, composting has shown efficiency in the sustainable management of solid waste wealthy in organic matter (carbon, nitrogen, and phosphorus, etc.) Furthermore, compost application to the soil allows the stimulation of microbial activity by rising microbial biomass activity, and therefore, their enzymatic reaction [5–8]. Composting is a stabilization and mineralization treatment of organic solid waste used to produce a stabilized compost [1,9,10]. A good management of composting process, from the feedstock to the final compost, is linked with optimum conditions.

Nonetheless, compost quality depends upon on its stability and maturity, and both must be monitored [4]. The intricacy of the composting process is assessed on the huge number of control variables (physical-chemical analysis, microbiological parameters, and enzymatic activities). These techniques are time consuming, and generally, more than one approach must be evaluated, which makes it difficult to assess the compost quality, thus favoring the exploration of new technologies for adequate assessing of the composting. Additionally, the maturity and stability of compost may also be achieved by spectroscopic analyses (X-ray diffraction, UV-Visible spectroscopy, infrared spectroscopy, scanning electron microscopy, etc.) These techniques are among the most reliable and promising methods for organic matter characterization as heterogeneous, thus providing complete information on functional and behavioral properties of samples [11].

In this regard, the use of spectroscopic methods has been suggested as a useful tool for assessing compost maturity. XRD analysis was usually used to investigate the organic matter transformation, especially that of cellulose, during composting. Infrared spectroscopy is a qualitative methods for assessing the organic matter functional groups [12]. Likewise, scanning electron microscopy (SEM) analysis was widely used to identify the morphological structure of compost that validates the final product stability and maturity. Recently, spectroscopic methods appear as a reliable tool for organic matter transformation prediction, particularly the decomposition of cellulose, during composting [11,13].

To date, no investigations are present which include both traditional and advanced methods to assess the maturity and stability of textile waste with high concentration (80%) during composting. The main goal of the current investigation was the establishment of an achievable technology to enhance the composting process monitoring of textile waste mixed with green waste, paper, and cardboard waste, using a rising amount of textile waste (80%), by monitoring its degree of maturity using physical-chemical parameters as well as infrared spectroscopy, scanning electron microscopy, and XRD on the one hand. On the other hand is to evaluate the feasibility of spectroscopy analysis to predict the composting treatment and to identify the correlation between quantitative and qualitative chemical properties through textile waste composting using PLS-R model.

## 2. Materials and Methods

### 2.1. Experimental Material

The compost produced in the current investigation was carried out using as technique aerobic silo composting. It was composed of textile waste (80%), green waste (10%), and paper & cardboard waste (10%). Textile waste is obtained from MULTIWACH textile industry located in Fez (Morocco). Whereas green waste, paper, and cardboard waste was collected from the Faculty of Sciences, Dhar El Mahraz, Fez. Main characteristics of feedstock were illustrated in Table 1. Firstly, these wastes were mixed and placed in silo composter of 200 L (size 0.58 × 0.58 × 0.92 m L × W × H) throughout 11 months. The silo composter was mixed and turned manually several times throughout these 11 months to promote organic matter humification. In order to have a representative and composite sample, samples were collected at different stages depending on three (0–20, 30–40, and 50–70 cm) and four necessary positions (north, south, east, and west).

**Table 1.** Feedstock physical-chemical characteristics.

| Physical-Chemical Parameters | Green Waste | Textile Waste | Paper and Cardboard Waste |
|---|---|---|---|
| Moisture% | 60.56 ± 0.04 | 50.26 ± 0.05 | 10.23 ± 1.06 |
| pH | 6.80 ± 0.40 | 7.30 ± 0.11 | 7.20 ± 0.01 |
| Total Organic Carbon% | 44.56 ± 0.60 | 30.26 ± 0.36 | 40.26 ± 0.04 |
| Total Kjeldahl Nitrogen% | 1.23 ± 1.00 | 0.53 ± 0.80 | 1.01 ± 0.20 |
| C/N Ratio | 36.22 ± 0.01 | 57.09 ± 0.15 | 39.86 ± 0.18 |

Values designate mean ± standard deviation based on 3 samples.

### 2.2. Experimental Analysis

#### 2.2.1. Physical-Chemical Analysis

Ascertainment of moisture, temperature, ash, and pH were performed according to the method depicted by the French Association for Standardization [14]. From the percentage of total organic carbon (TOC) and the total nitrogen (TN). The C/N ratio was calculated following the protocol provided by [15]. As for $Q_4/Q_6$ ($\lambda_{472}/\lambda_{664}$) absorbance ratio, analysis was performed according to the methods provided by [16]. Cellulose activity was measured according to [15]. Physical-chemical analysis was accomplished at various phases of composting (Initial, 3rd, 9th, and 11th months), these data of the same sample were published previously [15] (Table 2).

**Table 2.** Physical-Chemical properties changes throughout the composting process.

| Time (Months) | TOC (%) | TN (%) | C/N Ratio | $NH_4^+/NO_3^-$ | Ash (%) | Moisture (%) | Temperature (°C) | pH | Cellulose Activity (U $g^{-1}$) |
|---|---|---|---|---|---|---|---|---|---|
| **Initial** | 32.64 ± 0.96 | 0.56 ± 0.04 | 58.29 | 13.79 | 41.25 | 69.90 ± 3.00 | 10.50 ± 2.10 | 8.37 ± 0.10 | 0.41 |
| **3** | 28.26 ± 0.50 | 0.76 ± 0.01 | 37.18 | 6.09 | 49.13 | 34.55 ± 3.80 | 25.00 ± 2.00 | 7.53 ± 0.50 | 30.55 |
| **9** | 23.91 ± 0.45 | 1.17 ± 0.04 | 20.44 | 0.89 | 56.96 | 45.91 ± 2.30 | 42.00 ± 2.90 | 6.86 ± 0.10 | 42.59 |
| **11** | 23.53 ± 0.12 | 1.10 ± 0.02 | 21.39 | 0.28 | 57.65 | 37.24 ± 3.10 | 29.00 ± 1.20 | 6.02 ± 0.23 | 5.84 |

Values designate mean ± standard deviation based on 3 samples.

#### 2.2.2. X-ray Diffraction Analysis

X-ray diffraction (XRD) analysis was carried out using X'Pert Pro diffractometer using Cu K$\alpha$ monochromatic radiation ($\lambda$ = 1.5406 Å), with a generator of 30 mA and 40 kV. The 2$\theta$ angle was between 5 and 70° [17,18].

#### 2.2.3. Infrared Spectroscopy Analysis

For each sample, 2 mg of powder at various phases of composting (Ti: initial stage, T3: 3rd month, T9: 9th month, and T11: 11th month), were placed into a quartz-glass cell and analyzed with a IR-spectrometer (Bruker Vertex 70 spectrometer model) using a DTGS detector and OPUS 6.5 software, with wavenumber region between 400 and 4000 cm$^{-1}$ with 2 cm$^{-1}$ of resolution [12].

#### 2.2.4. Statistical Analysis

Statistics analyses of the physical-chemical fractions of biomaterial compost were achieved. Mean, standard deviation (St. Dev.), coefficient of variation (CV %), and a comparison of the means using the normalized analysis of variance (one-way ANOVA) were calculated. The PCA represent the correlation between chemical variables. Partial Least Squares-regression (PLS-R) was developed to correlate the chemical fractions of the compost samples with spectroscopy analyses. The PLS-R model is a multivariate linear calibration. In order to reduce the error margin, this model allows for minimizing large numbers of raw data into small sets of orthogonal factors among the values to be predicted.

## 3. Results and Discussion

### 3.1. Changes in Physical-Chemical Properties

The physical-chemical properties of mature compost were found to differ considerably compared to the beginning of treatment. Table 2 depicts that the temperature increase rapidly through progression of the composting process to reach a maximum of 42 °C compared to the initial value, this increase could be owing to the microorganisms activities and their metabolic reactions [19–21]. At the end of the treatment, a decrease in temperature was depicted, which might be owing to the predominance of molecules, which are difficult to biodegrade by microorganisms (lignocellulosic, etc.) Furthermore, a decrease in moisture was observed with a value of 34.24%, due to the evaporation of the water [15,22]. Regarding the pH, a significant decrease was recorded through composting treatment, and it was 6.02 in final compost, as equated to 8.37 in feedstock mixture, this could be ascribed to the organic acids generation during composting [15,23,24]. Hence, the pH tended toward neutral range at the end of the process. A considerable decrease was detected for total organic carbon (TOC) by 23.53% in final compost in comparison to feedstock mixture. This decrease might be assigned to the organic matter mineralization. The low level of TOC in final compost revealed a huge amount of humic acid, thus indicating a good stabilization and maturation of the final compost [12,23]. Moreover, an increase of total nitrogen (TN) was recorded in the final compost (1.10%) as contrasted to initial waste mixture (0.56%). Rise in TN value could be assigned to nitrogenous substrate mineralization performed through microorganisms activities [25]. The C/N ratio provides a general overview about compost stability and maturity levels. In the current study, C/N ratio considerably decreased to 21.39% in final compost as contrasted to the pre-composted mixture. The mineralization of carbon and the release of N during progression of the composting process led to decreased C/N ratio at the end of the process. Reducing in the C/N ratio might be assigned to the rise in humification rate and degradation of recalcitrant molecules such as lignin, hemicelluloses, and cellulose during composting [24]. Additionally, C/N ratio below 20 is considered a good index of compost stability [15], wherefore compost produced in the current investigation might be considered as stable and mature (C/N of 21.39%). The $NH_4^+/NO_3^-$ ratio is widely used to show the compost maturity. A $NH_4^+/NO_3^-$ ratio of 0.22 was recorded at the end of the process, which is an indicative of acceptable maturity according several authors [10,21]. Cellulase activity determines overall microbial activity and organic substrate stabilization through the composting process. This activity was recorded as 0.41 U $g^{-1}$ in initial feedstock mixture, which was increased to 42.59 U $g^{-1}$ at the middle of composting and declined to 5.84 U $g^{-1}$ in final compost. Rise in cellulose activity was assigned by several authors to microorganisms activities in the compost [15,23]. The substantial presence of cellulose activities towards the end of composting could be assigned to the formation of complexes between humic substances and cellulase enzymes, which is sheltered against denaturation and decaying [23]. Hence, it can be stated that composting of textile waste, even with high concentration of textile waste (80%), has tremendous potential for valorization of plant nutrients from these wastes.

The $Q_4/Q_6$ ($\lambda_{472}/\lambda_{664}$) ratio depicts the degree of aromatic constituents' polymerization and might be considered a humification index. At the end, a decrease in this ratio was detected with a value of 8.9 compared to the initial value (66.00) (Table 3). Moreover, stabilization of this ratio at the end of treatment reflects an increase in humic substance, depicted by phenolic and benzenecarboxylic compounds. According to several authors, a $Q_4/Q_6$ ratio around 5 might be referring to the humified matter, thus justifying a noteworthy decrease in the C/N ratio [16,26].

**Table 3.** Absorbance ratios evolution for the composting samples studied.

| Time (Months) | $Abs_{472nm}$ | $Abs_{664nm}$ | $Q_4/Q_6$ |
|---|---|---|---|
| **Initial** | 0.20 | 0.003 | 66.00 |
| **3** | 0.44 | 0.030 | 14.67 |
| **9** | 0.70 | 0.080 | 8.75 |
| **11** | 0.80 | 0.090 | 8.89 |

Values designate mean ± standard deviation based on 3 samples.

### *3.2. X-ray Diffraction Analysis*

Figure 1 depicts the XRD diffractograms for the compost samples at various composting stages (T3: 3rd month, T9: 9th month, and T11: 11th month) during textile waste composting. According X-ray diffractograms several peaks was depicted outspread over the range 2θ from 20° to 50°, thus allowing identification of the majority of crystalline fraction of cellulose intensities [27,28], with a noteworthy difference in intensity. X-ray diffractograms of samples analysis, indicating the attendance of the peaks attributed to the cellulose I fraction (2θ = 26.56°) and to cellulose II (2θ = 34.50°). In comparison to initial stage, the XRD spectra depicted significant lessening in the peak intensity at the end of composting, reflecting the decomposition of the cellulose polymer chains caused mainly by microorganisms through their enzymatic system, thus justifying the C/N ratios reduction and therefore the increasing in the cellulose activity during composting, reflecting the cellulose compounds degradation [18,27]. Equally, peaks at 2θ = 36.50°, 40.25°, and 44.64°, attributed to calcium carbonate $CaCO_3$ (inorganic fraction) was observed, thus confirming that there was an organic matter mineralization into inorganic constituents by microorganisms during composting.

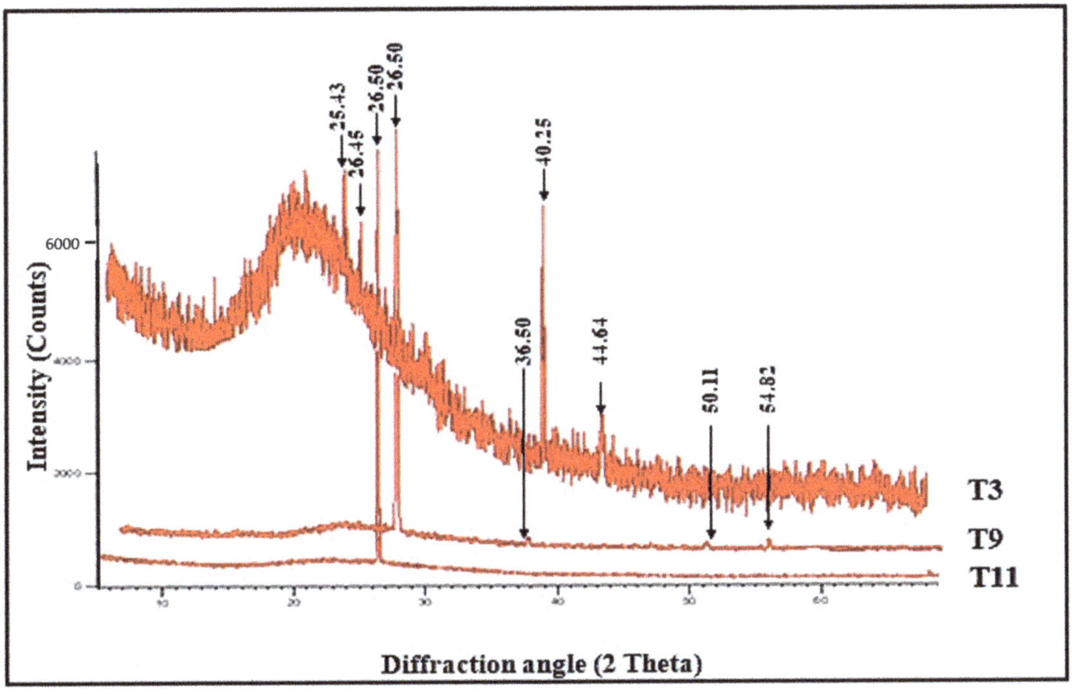


**Figure 1.** X-ray diffractogram of mixture at different stages of composting (T3: 3rd month, T9: 9th month and T11: 11th month).

### 3.3. Infrared Spectroscopy Analysis

The chemical functional groups of organic matter was identified using infrared spectroscopy [18,23]. In Figure 2, the pertinent spectral bands of feedstock mix and final compost are depicted. The infrared (IR) absorptions assignment is summarized in Table 4. Spectra present in Figure 2 confirmed the presence of crystalline cellulose, as well as hemicellulose and lignin, through of the bands recorded in the range 3700–3100 cm$^{-1}$. In comparison to feedstock mixture, final compost spectra depicted a considerable decreasing in the intensity of the peak at 3348 cm$^{-1}$, thus indicating decaying by microorganisms of these compounds [1,9,29,30]. The band between 1700–1600 cm$^{-1}$ owing to N-H bending vibrations attributed to Amide I in hemicelluloses and/or cellulosic in compost samples (Figure 2) [30]. The presence of mineral form was corroborated by the peak at 1399 cm$^{-1}$, thus justifying the results obtained previously by XRD and Ash (Table 2). According to [12], cellulose in crystallized form is depicted in infrared spectrum by a small band at 1348 cm$^{-1}$, thus proving the XRD results mentioned beforehand (Figure 1). The band appeared at 1120 cm$^{-1}$, indicating the presence of lignin and carbohydrates [31,32]. Band between 1172 and 950 cm$^{-1}$ was detected, which depicts the fingerprints region of cellulose and hemicelluloses. Additionally, a peak at 1030 cm$^{-1}$ owing to aromatic compound and carbohydrates in compost samples was observed [27,33–35]. A prominent increase in this band was noticed, owing to raise levels of aromatic compound and polysaccharides [36]. These findings proved the decomposition of polysaccharides, aliphatic, and carbohydrates from the feedstock mixture, which corroborate that composting induced the cellulose degradation, whereas humic structures and aromatic compounds appeared towards the end of composting, showing its stability and maturity, and equally corroborate the results obtained by the physical-chemical analysis.

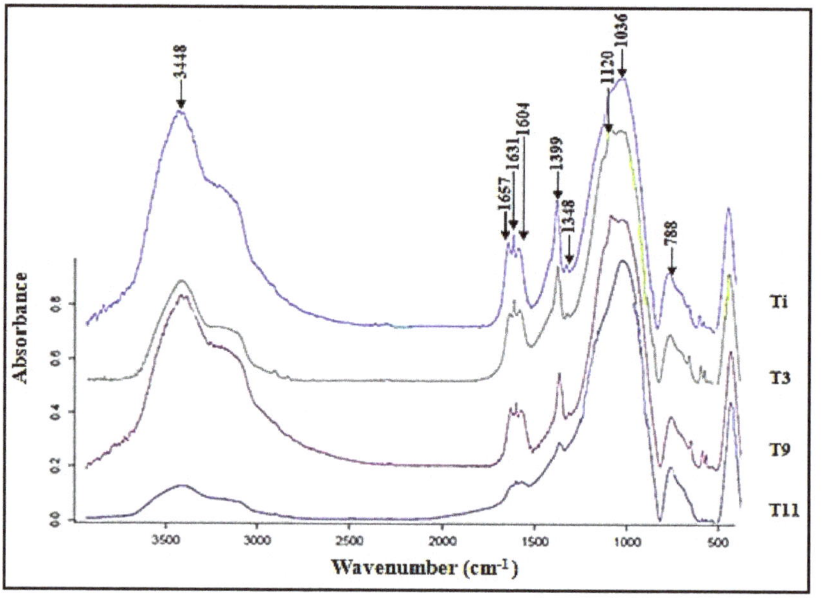


**Figure 2.** Infrared spectra indicating the evolution of compost samples at various stages of composting (T3: 3rd month, T9: 9th month and T11: 11th month).

**Table 4.** Assignments of the main vibrations of the Infrared spectra.

| Wavenumber ($cm^{-1}$) | Band Assignements |
|---|---|
| 3700–3000 | v(OH) hydroxyl groups in lignin, cellulose and hemicelluloses Intermolecular hydrogen-bonded |
| 1700–1600 | vC=O ester in acetoxy groups (H3C–(C=O) –O–) in hemicellulose, vC=O in quinone or p-quinone |
| 1400–1300 | δC–H and δsCH3 in cellulose and hemicelluloses |
| 1120 | CH stretching vibrations in different groups of lignin and cellulose and hemicelluloses |
| 1030 | C–O–C stretching vibration of lignin and polysaccharides |
| 787 | CH2 rocking vibration in cellulose (cellulose $I_\beta$ appear generally as a tiny peak at 750 $cm^{-1}$ and 3240 $cm^{-1}$) |

The assignments and chemicals are based on [12,17,18].

### 3.4. Scanning Electron Microscopy (SEM)

Observations performed using scanning electron microscopy (SEM) allowed a gain of good knowledge of feedstock and final compost typical morphology. The morphological properties surface of compost samples were studied using SEM under 2000 magnification and illustrated in Figure 3a,b.

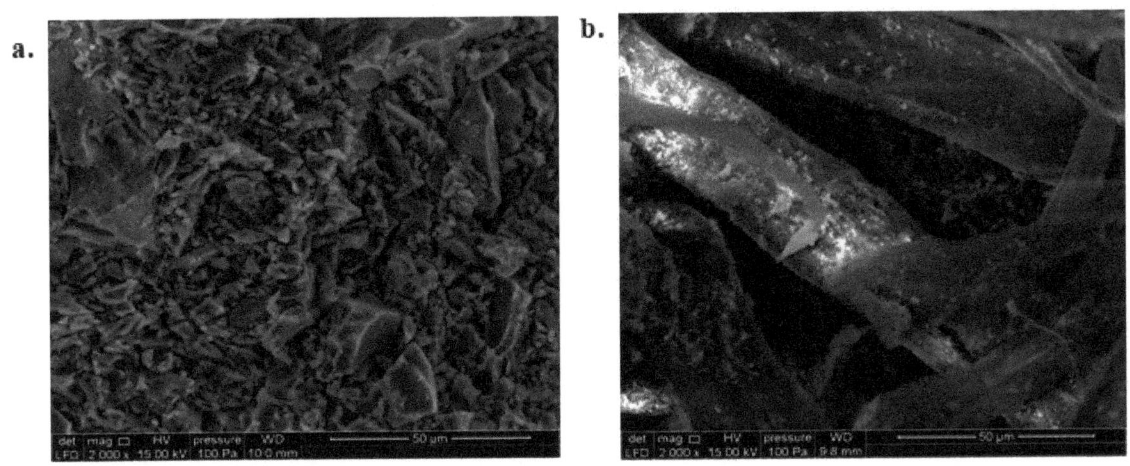


**Figure 3.** Scanning electron microscopy (2000×) of (**a**). pre-composted waste mixture and; (**b**). final compost.

The SEM Snapshots recorded of feedstock and final compost are compared in Figure 3a,b. The analysis depicted that the structure of the feedstock mixture was considerably compacted and flock, whilst a more granular, porous, and fragmented structure was depicted for the final compost (Figure 3a,b). Composting is a biological treatment based mainly on microorganism activity and their enzymatic systems. The disintegration of the substrate during composting was facilitated by various enzymes secreted (especially cellulose activity, etc.) by the microorganisms in the compost. Through Figure 3b, it can be noticed that final compost had fibrous structure; these fibers have different sizes and shapes, with diameters ranging between 5 and 45 µm, which corresponds, presumably, to cellulose fibers. The whole chains of fibers bundle together randomly, contributing to a heterogeneous network structure. These findings are consistent with various authors who used SEM analysis to highlight the modification that occurred in surface morphology of initial and final substrate composted [23,32,37].

### 3.5. Statistical Analysis

The factorial map depicted mixture samples of each age garnered showed by barycenter co-ordinates. The PC1 and PC2 reflected, respectively, 81.40% and 9.41% of the variance in spectral data (Figure 4). Using the PCA analysis, the chemical parameter of composts changed according to the composting stage. A chronological apportionment of samples on PC2 was observed. C, N, C/N ratio, $NH_4^+/NO_3^-$ ratio, $Q_4/Q_6$ ratio, cellulose activity, pH, T°, and Moisture arrows, pointed from old samples (44-week-old) to young composts (1-week-old). This finding could explain the change in C/N ratio throughout composting. Several authors explain that the modification in C/N ratio indicates the organic matter stabilization and decay [7,12,38]. Arrow of C/N, directed from old to young compost groups, shows the organic matter stabilization and strengthens the chronological distribution on PC2, and comparable outcomes were noticed by [4].

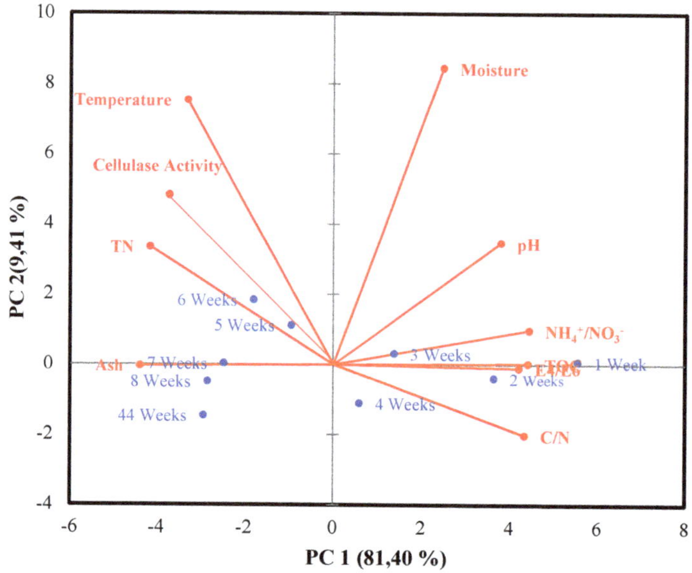


**Figure 4.** Results of PCA analysis. Each circle is the barycentre co-ordinate with standard deviation of each stage of composting. TOC: total organic carbon; TN: total nitrogen; C/N: carbon/nitrogen ratio; $NH_4^+/NO_3^-$ ratio: ammonium nitrogen /nitrate nitrogen; $Q_4/Q_6$ ratio: 472/664 nm.

The main goal of the current investigation was to indicate the compost development of compost from a high concentration of textile waste (80%). For this reason, the prediction of organic matter state depending upon time using infrared analysis to predicate C, N, C/N, $NH_4^+/NO_3^-$, $Q_4/Q_6$, Cellulose activity, T°, time of composting, and moisture was employed. The accurateness of the prediction model was estimated by several parameters. The $R^2$ values were 0.98, 0.98, 0.98, 0.97, 0.79, 0.98, 0.77, 0.37, and 0.90, respectively for C, N, C/N, $NH_4^+/NO_3^-$, $Q_4/Q_6$ ratio, cellulose activity, T°, moisture and composting time validates the model (Table 5), which prove that these parameters are well correlated with IR spectra. The RMSEP values were 0.58, 0.07, 3.10, 0.64, 5.02, 1.26, 3.89, 7.88, and 0.90, respectively for C, N, C/N, $NH_4^+/NO_3^-$, $Q_4/Q_6$ ratio, T°, moisture, and time of composting, which is acceptable according several authors [38,39]. Moreover, comparable outcomes were noticed by [38,39]. Nonetheless, more selective criteria, such as the coefficient of variation and standard deviation, to assess the efficiency of infrared spectroscopy calibrations were used. Regarding the coefficient of variation (CV%), the values were of 2.29, 7.87, 0.52, 12.74, 21.91, 46.49, 17.44, 17.76, and 33.04, respectively for C, N, C/N, $NH_4^+/NO_3^-$, $Q_4/Q_6$ ratio, T°, moisture, and time of composting. ANOVA test was used to depict the changes of the physical-chemical properties during textile waste composting. Significant differences ($p$-value < 0.05) were recorded for all physical–chemical parameters except moisture.

**Table 5.** Summary statistics of compost analytical parameters and Infrared spectra.

| | Mean | St. Dev. | CV (%) | ANOVA ($p$) | $R^2$ | RMSEP * |
|---|---|---|---|---|---|---|
| **Time** | 5.75 | 1.90 | 33.04 | <0.05 | 0.90 | 1.35 |
| **TOC** | 27.08 | 0.62 | 2.29 | <0.05 | 0.98 | 0.58 |
| **TN** | 0.89 | 0.07 | 7.87 | 0.004 | 0.98 | 0.07 |
| **C/N** | 34.32 | 3.28 | 0.82 | 0.002 | 0.98 | 3.10 |
| **$NH_4^+/NO_3^-$** | 5.26 | 0.67 | 12.74 | <0.05 | 0.97 | 0.64 |
| **T°** | 23.62 | 4.12 | 17.44 | 0.045 | 0.77 | 3.89 |
| **Ph** | 7.19 | 0.40 | 5.56 | <0.05 | 0.68 | 0.38 |
| **Moisture** | 46.90 | 8.33 | 17.76 | 0.143 ns | 0.37 | 7.88 |
| **$Q_4/Q_6$** | 24.23 | 5.31 | 21.91 | 0.021 | 0.79 | 5.02 |
| **Cellulase activity** | 19.85 | 9.23 | 46.49 | <0.05 | 0.98 | 1.26 |

CV: coefficient of variation; Significance levels: ns: not significant; $p < 0.05$; $R^2$: coefficient of correlation; RMSEP *: root mean square error of prediction; TOC: total organic carbon; TKN: total Kjeldahl nitrogen; C/N: carbon/nitrogen ratio; $NH_4^+/NO_3^-$: ammonium nitrogen /nitrate nitrogen; $Q_4/Q_6$ ratio: 472/664 nm.

Through this model, the differentiation between the properties of the compost, the phases of organic matter transformation could be equally highlighted. Actually, it was demonstrated that this analysis may use it to predict C/N and TN development in compost samples [38,39]. Nitrogen is considered the most important constituents in fertilizers, and, equally, it is a very important element present in textile, paper and cardboard, and green waste. At the beginning, a part of the nitrogen content in composted wastes has undergone volatilization into air during the composting process [40]. At the next phase, organic N was transformed to ammonium and nitrate-N by microorganisms, thus raising the amount of nitrate-N, making it effective as a nitrogen fertilizer. The capacity to predict TN concentration in compost samples was strong and accurate. Moreover, C/N ratio is widely used as a good and reliable tool to determine the maturity of compost; the PLS-R model was more successful for the compost samples. Additionally, $Q_4/Q_6$ ratio is considered, by several authors, as a good humification index, the predictions of this parameter was strong (Table 5) [12,38,39]. Commonly, the restoring of the predicted results depending upon the time depicted that no differences recorded between the measured and predicted values for the majority of the parameters (Figure 5). This investigation strengthened the chronological apportionment of compost provided by PCA following an advance of composting maturity. Another advantage of this investigation was to predict time of composting using different composts samples.

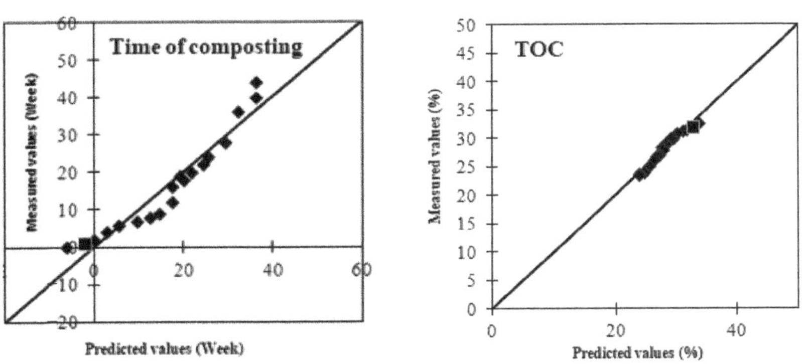


**Figure 5.** *Cont.*

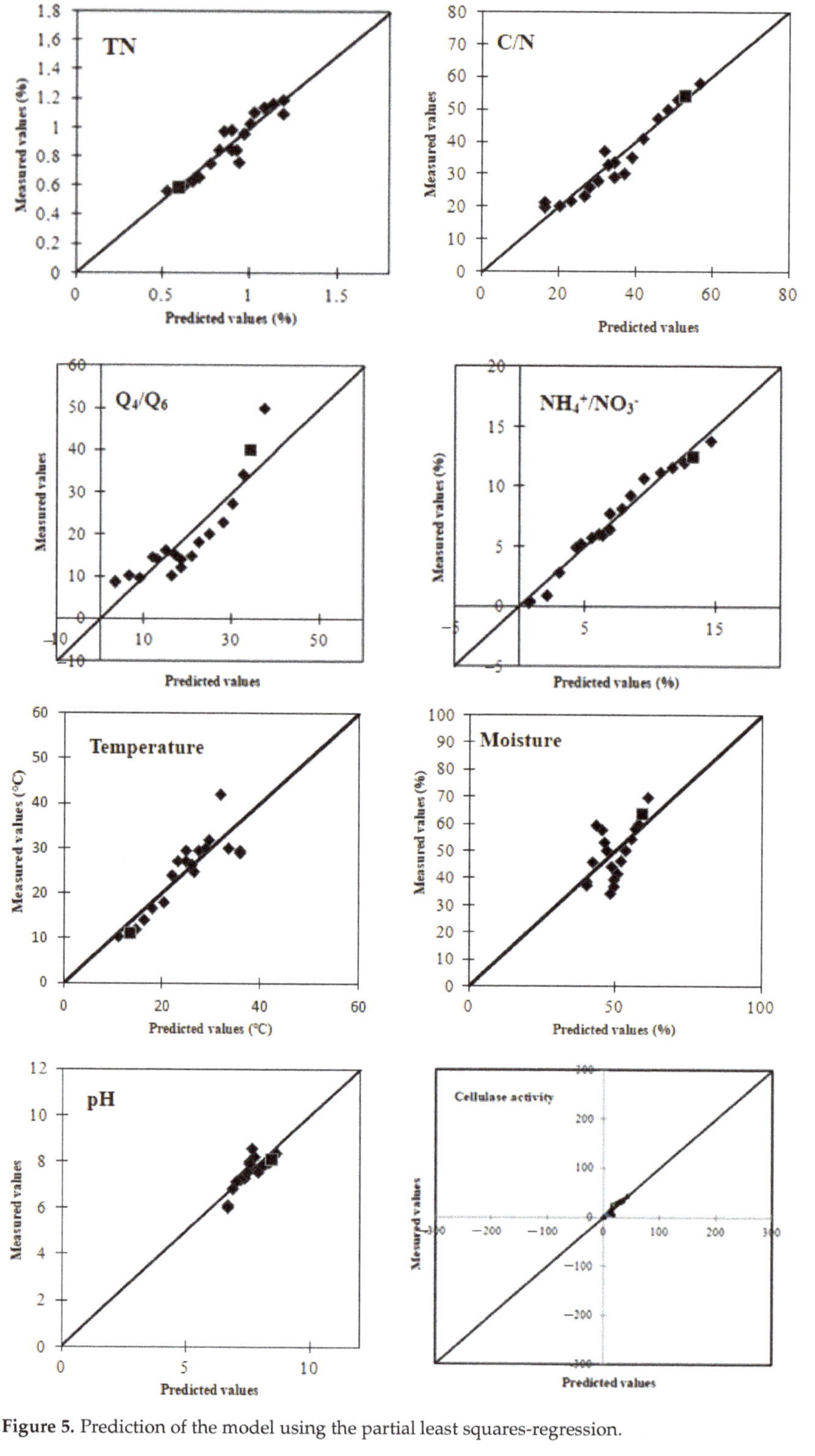


**Figure 5.** Prediction of the model using the partial least squares-regression.

## 4. Conclusions

This study aimed to evaluate the applicability of spectral techniques to characterize the compost quality from textile waste even with a high concentration of this waste (80%). For this reason, quantitative and qualitative physical-chemical parameters of compost samples were correlated with spectral analysis. A decrease in aliphatic compounds was observed in contrast to aromatic compounds, throughout solid waste decaying corroborated by XRD analysis. In addition, SEM micrographs depicted relatively larger areas and a plenty fibrous structure of final compost, thus proving cellulose fibers decaying. The current study provides an advance technique to prove compost maturity and stability from a huge concentration of textile waste. Ultimately, through this investigation, it can be concluded that the compost produces meet the required level of maturity.

**Author Contributions:** Conceptualization, S.B. and M.M.; methodology, S.B., M.M., H.I., H.M., K.E.; validation, M.M. and S.B. Data curation, S.B.; Formal analysis, S.B.; Investigation, S.B.; writing—original draft preparation, S.B.; Review and editing S.B., M.M., M.F.A. and R.C. Supervision, M.M. and M.B.; Funding acquisition, M.F.A., R.C. All authors have read and agreed to the published version of the manuscript.

**Funding:** This research was funded by Researchers Supporting Project number RSP-2021/122, King Saud University, Riyadh, Saudi Arabia.

**Institutional Review Board Statement:** Not applicable.

**Informed Consent Statement:** Not applicable.

**Data Availability Statement:** The data presented in this study are available on request from the corresponding author.

**Acknowledgments:** The authors acknowledge the generous support from researchers supporting project-under the group number (RSP-2021/122), King Saud University, Riyadh, Saudi Arabia; and the National Center for Scientific Research in Rabat (CNRST). The authors are gratefully acknowledged to BOUKIR Abdellatif from Faculty of Sciences and Techniques Fez for support and help.

**Conflicts of Interest:** The authors declare no conflict of interest.

## References


1. Zahra El Ouaqoudi, F.; El Fels, L.; Lemée, L.; Amblès, A.; Hafidi, M. Evaluation of Lignocelullose Compost Stability and Maturity Using Spectroscopic (FTIR) and Thermal (TGA/TDA) Analysis. *Ecol. Eng.* **2015**, *75*, 217–222. [CrossRef]
2. Talouizte, H.; Merzouki, M.; Benlemlih, M. Treatment of Real Textile Wastwater Using SBR Technology: Effect of Sludge Age and Operational Parameters. *J. Biotechnol. Lett.* **2017**, *4*, 79–83.
3. Pagliai, M.; Vignozzi, N.; Pellegrini, S. Soil Structure and the Effect of Management Practices. *Soil Tillage Res.* **2004**, *79*, 131–143. [CrossRef]
4. Albrecht, R.; Joffre, R.; Gros, R.; Le Petit, J.; Terrom, G.; Périssol, C. Efficiency of Near-Infrared Reflectance Spectroscopy to Assess and Predict the Stage of Transformation of Organic Matter in the Composting Process. *Bioresour. Technol.* **2008**, *99*, 448–455. [CrossRef]
5. Wu, H.; Zhao, Y.; Long, Y.; Zhu, Y.; Wang, H.; Lu, W. Evaluation of the Biological Stability of Waste during Landfill Stabilization by Thermogravimetric Analysis and Fourier Transform Infrared Spectroscopy. *Bioresour. Technol.* **2011**, *102*, 9403–9408. [CrossRef]
6. Albrecht, R.; Le Petit, J.; Calvert, V.; Terrom, G.; Périssol, C. Changes in the Level of Alkaline and Acid Phosphatase Activities during Green Wastes and Sewage Sludge Co-Composting. *Bioresour. Technol.* **2010**, *101*, 228–233. [CrossRef]
7. Crecchio, C.; Curci, M.; Pizzigallo, M.D.R.; Ricciuti, P.; Ruggiero, P. Effects of Municipal Solid Waste Compost Amendments on Soil Enzyme Activities and Bacterial Genetic Diversity. *Soil Biol. Biochem.* **2004**, *36*, 1595–1605. [CrossRef]
8. Peláez, C.; Mejía, A.; Planas, A. Development of a Solid Phase Kinetic Assay for Determination of Enzyme Activities during Composting. *Process Biochem.* **2004**, *39*, 971–975. [CrossRef]
9. Vergnoux, A.; Guiliano, M.; Le Dréau, Y.; Kister, J.; Dupuy, N.; Doumenq, P. Monitoring of the Evolution of an Industrial Compost and Prediction of Some Compost Properties by NIR Spectroscopy. *Sci. Total Environ.* **2009**, *407*, 2390–2403. [CrossRef]
10. Aguelmous, A.; Lahsaini, S.; El Fels, L.; Souabi, S.; Zamama, M.; Hafidi, M. Biodegradation Assessment of Biological Oil Sludge from a Petroleum Refinery. *J. Mater. Environ. Sci.* **2016**, *7*, 3421–3430.
11. Temporal-lara, B.; Melendez-pastor, I.; Gómez, I.; Navarro-pedreño, J. Wastewater Biosolid Composting Optimization Based on UV-VNIR Spectroscopy Monitoring. *Sensors* **2016**, *16*, 1919. [CrossRef]
12. Ilani, T.; Herrmann, I.; Karnieli, A.; Arye, G. Characterization of the Biosolids Composting Process by Hyperspectral Analysis. *Waste Manag.* **2016**, *48*, 106–114. [CrossRef]

13. Ueno, M.; Taira, E.; Kawamitsu, Y.; Komiya, Y.; Kikuchi, K. Application of FT-NIR Spectroscopy to the Evaluation of Compost Quality. *Eng. Agric. Environ. Food* **2008**, *1*, 51–56. [CrossRef]
14. Afnor. *Amendements du sol et Support de Culture-Préparation des Echantillons Pour les Essais Physiques et Chimiques, Détermination de la Teneur en Matière Sèche, du Taux D'humidité et de la Masse Volumique Compactée en Laboratoire*; The French Association for Standardization: Paris, France, 2000.
15. Biyada, S.; Merzouki, M.; Demcenko, T.; Vasiliauskiene, D.; Urbonavicius, J.; Marciulaitiene, E.; Vasarevicius, S.; Benlemlih, M. Evolution of Microbial Composition and Enzymatic Activities during the Composting of Textile Waste. *Appl. Sci.* **2020**, *10*, 3758. [CrossRef]
16. Albrecht, R.; Le Petit, J.; Terrom, G.; Périssol, C. Comparison between UV Spectroscopy and Nirs to Assess Humification Process during Sewage Sludge and Green Wastes Co-Composting. *Bioresour. Technol.* **2011**, *102*, 4495–4500. [CrossRef] [PubMed]
17. Boukir, A.; Mehyaoui, I.; Fellak, S.; Asia, L.; Doumenq, P. The Effect of the Natural Degradation Process on the Cellulose Structure of Moroccan Hardwood Fiber: A Survey on Spectroscopy and Structural Properties. *Mediterr. J. Chem.* **2019**, *8*, 179–190. [CrossRef]
18. Boukir, A.; Fellak, S.; Doumenq, P. Structural Characterization of Argania Spinosa Moroccan Wooden Artifacts during Natural Degradation Progress Using Infrared Spectroscopy (ATR-FTIR) and X-ray Diffraction (XRD). *Heliyon* **2019**, *5*, e02477. [CrossRef]
19. Wang, K.; Li, W.; Gong, X.; Li, Y.; Wu, C.; Ren, N. Spectral Study of Dissolved Organic Matter in Biosolid during the Composting Process Using Inorganic Bulking Agent: UV e Vis, GPC, FTIR and EEM. *Int. Biodeterior. Biodegrad.* **2013**, *85*, 617–623. [CrossRef]
20. Abdel-Rahman, M.A.; Nour El-Din, M.; Refaat, B.M.; Abdel-Shakour, E.H.; Ewais, E.E.D.; Alrefaey, H.M.A. Biotechnological Application of Thermotolerant Cellulose-Decomposing Bacteria in Composting of Rice Straw. *Ann. Agric. Sci.* **2016**, *61*, 135–143. [CrossRef]
21. Hachicha, R.; Hachicha, S.; Trabelsi, I.; Woodward, S.; Mechichi, T. Evolution of the Fatty Fraction during Co-Composting of Olive Oil Industry Wastes with Animal Manure: Maturity Assessment of the End Product. *Chemosphere* **2009**, *75*, 1382–1386. [CrossRef]
22. López-gonzález, J.A.; Vargas-garcía, M.C.; López, M.J.; Suárez-estrella, F.; Jurado, M.; Moreno, J. Biodiversity and Succession of Mycobiota Associated to Agricultural Lignocellulosic Waste-Based Composting. *Bioresour. Technol.* **2015**, *187*, 305–313. [CrossRef]
23. Srivastava, V.; Goel, G.; Kumar, V.; Pratap, R.; De Araujo, A.S.F.; Singh, P. Analysis and Advanced Characterization of Municipal Solid Waste Vermicompost Maturity for a Green Environment. *J. Environ. Manag.* **2020**, *255*, 109914. [CrossRef]
24. Soobhany, N.; Gunasee, S.; Pooja, Y.; Joyram, H.; Raghoo, P. Spectroscopic, Thermogravimetric and Structural Characterization Analyses for Comparing Municipal Solid Waste Composts and Vermicomposts Stability and Maturity. *Bioresour. Technol.* **2017**, *236*, 11–19. [CrossRef]
25. Sharma, A.; Ganguly, R.; Kumar, A. Spectral Characterization and Quality Assessment of Organic Compost for Agricultural Purposes. *Int. J. Recycl. Org. Waste Agric.* **2019**, *8*, 197–213. [CrossRef]
26. Amir, S.; Hafidi, M.; Merlina, G.; Revel, J.C. Structural Characterization of Fulvic Acids during Composting of Sewage Sludge. *Process Biochem.* **2005**, *40*, 1693–1700. [CrossRef]
27. Hajji, L.; Boukir, A.; Assouik, J.; Luis, J.; Luisa, M. Artificial Aging Paper to Assess Long-Term Effects of Conservative Treatment. Monitoring by Infrared Spectroscopy (ATR-FTIR), X-ray Diffraction (XRD), and Energy Dispersive X-ray Fl Uorescence (EDXRF) ☆. *Microchem. J.* **2016**, *124*, 646–656. [CrossRef]
28. Hajji, L.; Boukir, A.; Assouik, J.; Lakhiari, H.; Kerbal, A.; Doumenq, P.; Mille, G.; Luisa, M.; Carvalho, D. Conservation of Moroccan Manuscript Papers Aged 150, 200 and 800 Years. Analysis by Infrared Spectroscopy (ATR-FTIR), X-ray Diffraction (XRD), and Scanning Electron Microscopy Energy Dispersive Spectrometry (SEM-EDS). *Spectrochim. Acta Part A Mol. Biomol. Spectrosc.* **2015**, *136*, 1038–1046. [CrossRef]
29. Lim, S.L.; Wu, T.Y. Determination of Maturity in the Vermicompost Produced from Palm Oil Mill Effluent Using Spectroscopy, Structural Characterization and Thermogravimetric Analysis. *Ecol. Eng.* **2015**, *84*, 515–519. [CrossRef]
30. Garriga, P.; Colom, X.; Carrillo, F.; Nogue, F. Structural Analysis of Photodegraded Wood by Means of FTIR Spectroscopy. *Polym. Degrad. Stab.* **2003**, *80*, 543–549. [CrossRef]
31. Zghari, B.; Doumenq, P.; Romane, A.; Boukir, A. GC-MS, FTIR and 1 H, 13 C NMR Structural Analysis and Identification of Phenolic Compounds in Olive Mill Wastewater Extracted from Oued Oussefrou Effluent (Beni Mellal-Morocco). *J. Mater. Environ. Sci.* **2017**, *8*, 4496–4509.
32. Hajji, L.; Boukir, A.; Assouik, J.; Kerbal, A.; Kajjout, M. A Multi-Analytical Approach for the Evaluation of the Efficiency of the Conservation—Restoration Treatment of Moroccan Historical Manuscripts Dating to the 16th, 17th, and 18th Centuries. *Appl. Spectrosc.* **2015**, *69*, 920–938. [CrossRef] [PubMed]
33. Makarem, M.; Lee, C.M.; Kafle, K.; Huang, S.; Chae, I.; Yang, H.; Kubicki, J.D.; Kim, S.H. *Probing Cellulose Vibrational Spectroscopy*; Springer: Dordrecht, The Netherlands, 2019; ISBN 1057001821.
34. Acharya, S.; Hu, Y.; Moussa, H.; Abidi, N. Preparation and Characterization of Transparent Cellulose Films Using an Improved Cellulose Dissolution Process. *J. Appl. Polym. Sci.* **2017**, *44871*, 1–12. [CrossRef]
35. Falcão, L.; Araújo, M.E.M. Application of ATR—FTIR Spectroscopy to the Analysis of Tannins in Historic Leathers: The Case Study of the Upholstery from the 19th Century Portuguese Royal Train. *Vib. Spectrosc.* **2014**, *74*, 98–103. [CrossRef]
36. Droussi, Z.; Valeria, D.; Rosaria, M.; Hafidi, M.; Ouatmane, A. Study of the Biodegradation and Transformation of Olive-Mill Residues during Composting Using FTIR Spectroscopy and Differential Scanning Calorimetry. *J. Hazard. Mater.* **2009**, *164*, 1281–1285. [CrossRef]

37. Katusiimeh, M.W.; Burger, K.; Mol, A.P.J. Informal Waste Collection and Its Co-Existence with the Formal Waste Sector: The Case of Kampala, Uganda. *Habitat Int.* **2013**, *38*, 1–9. [CrossRef]
38. Chang, C.; Laird, D.A. Near-Infrared Reflectance Spectroscopic Analysis of Soil c and n. *Soil Sci.* **2002**, *167*, 110–116. [CrossRef]
39. Malley, D.F.; Mcclure, C.; Martin, P.D. Compositional Analysis of Cattle Manure During Composting Using a Field—Portable Near—Infrared Spectrometer. *Commun. Soil Sci. Plant Anal.* **2005**, *36*, 455–475. [CrossRef]
40. Jouraiphy, A.; Amir, S.; El Gharous, M.; Revel, J.C.; Hafidi, M. Chemical and Spectroscopic Analysis of Organic Matter Transformation during Composting of Sewage Sludge and Green Plant Waste. *Int. Biodeterior. Biodegrad.* **2005**, *56*, 101–108. [CrossRef]

processes

MDPI

Article

# Agronomic Characteristics of the Compost-Bedded Pack Made with Forest Biomass or Sawdust

Lourdes Llonch [1], Cecilia Gordo [2], Marga López [2], Lorena Castillejos [1,*], Alfred Ferret [1] and Teresa Balanyà [2]

[1] Animal Nutrition and Welfare Service (SNiBA), Department of Animal and Food Science, Universitat Autònoma de Barcelona, 08193 Bellaterra, Spain; Lourdes.Llonch@uab.cat (L.L.); Alfred.Ferret@uab.cat (A.F.)

[2] Department of Agrofood Engineering and Biothecnology, School of Agro-Food and Biosystems Engineering, Universitat Politècnica de Catalunya, 08860 Castelldefels, Spain; ceci-2111@hotmail.com (C.G.); marga.lopez@upc.edu (M.L.); teresa.balanya@upc.edu (T.B.)

* Correspondence: Lorena.Castillejos@uab.cat

**Abstract:** To ascertain the agronomic value of the material resulting from the compost-bedded pack (CBP) in dairy barns, a cross-over experiment was designed with eight dry non-pregnant Holstein cows. The study was performed in two 11-week periods. Bedding materials used were: (1) CBP with sawdust (S) and (2) CBP with forest biomass (FB). Samples were taken from the raw bedding materials and from the CBP across the experiment. We conducted an additional study preparing two piles, one of each CBP material, to accomplish a composting process of 3 months, where samples were also taken. Granulometry and some chemical composition characteristics of FB made it a suitable bedding material to be used as CBP, but its high moisture content limited the ability to absorb liquid manure. Both the degree of stability of the organic matter and the temperature evolution of CBP suggest that a real composting process did not occur. Finally, the composting process of the piles did not lead to any relevant change in CBP materials. From the agronomic point of view, S and FB present potentially valuable characteristics as regards organic amendment in the soil, thanks to their high organic matter content and low nutrient content.

**Keywords:** compost-bedded pack; dairy cows; forest biomass; organic amendment; sawdust

**Citation:** Llonch, L.; Gordo, C.; López, M.; Castillejos, L.; Ferret, A.; Balanyà, T. Agronomic Characteristics of the Compost-Bedded Pack Made with Forest Biomass or Sawdust. *Processes* **2021**, *9*, 546. https://doi.org/10.3390/pr9030546

Academic Editor: Antoni Sánchez

Received: 2 March 2021
Accepted: 16 March 2021
Published: 19 March 2021



## 1. Introduction

Fifty-five percent of the world's population lives in urban areas, a proportion that is expected to increase to 68% by 2050 [1]. In this process, the number of both farms and farmers has decreased, thus raising, at the same time, the urban population [2]. One of the main consequences of this rural flight is an increase in forestland. Human activities have directly caused approximately 60% of the new global tree growth [3]. Dry weather and damaged ecosystems with an accumulation of dead biomass due to rural abandonment increase the risk of forest fires. The accumulation of high fuel loads over large areas is the main reason behind the occurrence of large fires [4,5]. In this context, it is necessary to find new ways to reduce these high fuel loads to prevent large fires.

The compost-bedded pack (CBP) is a loose housing system that is becoming more and more established on dairy farms in which cows are maintained for a long time on the compost. Improvements in the health, welfare and performance of cows, ease of farm chores and reduced building costs have been described in comparison with other housing systems [6–8]. The composting process allows manure and urine to be stored, as long as the pack is managed adequately, involving twice-daily tilling and periodic bedding addition [7,9,10]. Tilling incorporates manure and air into the pack, thus promoting aerobic microbiological activity, heating the pack and drying the lying surface for cattle to lie on [11]. Bedding addition increases the water-holding capacity of the pack to control CBP moisture. Compost-bedded pack barns in dairy cow farming mainly use sawdust (S) as a bedding material, but forest biomass (FB) could be an alternative material. When both materials were tested as CBP, some important microbial species affecting cow health were better controlled using FB, although in terms of cow comfort, FB did not appear to work as well as S [12]. However, the welfare assessment conducted did not detect health problems in the cows [12].

Once the CBP material is moved away from the barn, it is used as organic fertilizer on the farm. The lack of information about the agronomic value of this fertilizer source led us to study (a) the agronomic characteristics of the CBP obtained either with FB or S, after maintaining cows for three months on the pack, and (b) the agronomic characteristics of both composted materials after conducting an additional composting process of three months once the cows were moved away.

## 2. Materials and Methods

Two different composting processes were established. The first involved maintaining cows on CBP, either FB or S, for 11 weeks. Once this one was completed, a second process began, based on composting piles from the two CBPs for 3 months.

### 2.1. Bedding and Compost Management

Eight dry non-pregnant Holstein cows were individually allocated in roofed concrete floor pens, 5 m long and 2.5 m wide. Each pen was divided in 2 areas: a feeding area equipped with a feed bunk and a water trough, and a resting area. Pens were separated by a metal fence that allowed contact between animals. Cows were randomly assigned to 1 of 2 treatments in a cross-over design with 4 cows per treatment. The study was performed in 2 11-week periods with a 4-week washout period between them. Bedding materials used were: (1) CBP with S (CBP-S) and (2) CBP with FB (CBP-FB). The bulk density for these bedding materials was 182 $\pm$ 6.3 g/L and 240 $\pm$ 16.2 g/L for S and FB, respectively. During the washout period, cows laid on traditional wood shaving bedding. At the beginning of each period, pens were filled with 30 cm of the new bedding material. All CBPs were tilled twice daily (10 a.m. and 5 p.m.) at 30 cm depth with a rototiller, and an average of 0.8 kg/m$^2$/d of new bedding material was added on each pen CBP surface before tilling, when pen CBP moisture was greater than 60%. Compost-bedded pack moisture was measured weekly in both areas (feeding and resting) of each pen. Averages of 7.8 kg/pen/d in CBP-S and 7.9 kg/pen/d in CBP-FB of new bedding materials were added. Both CBPs were removed at the end of week 11 of each period. Daily ambient temperature and environmental humidity were obtained from 2 data loggers (UX100-003, Hobo, Algete-Madrid, Spain) located in the barn. At the end of the second experimental period, the composted bedding material of each pen was composited and used to prepare 2 composting piles, one of each bedding material, to obtain a composted material after 3 months of the composting process. Each pile, in the shape of a truncated pyramid and located in a roofed barn, was 4.8 m long, 2.5 m wide and 1.15 m high, with an approximate volume of 7.5 m$^3$. The piles were turned and rebuilt weekly to facilitate the aerobic process. It was necessary to water the piles on five occasions to maintain pile moisture between 45 and 65%. Daily ambient temperature and environmental humidity were obtained as described before.

### 2.2. Sampling

Samples were taken from the raw bedding materials, the CBP and the composted piles. Raw bedding material was sampled throughout the experiment. Thus, in total, 4 samples of each one were collected and then stored at 4 °C until analysis. Particle size of the raw bedding materials was measured using an electromagnetic sieve shaker (RP 200N, CISA Cedaceria Industrial S. L., Barcelona, Spain) to obtain their physical characterization. With regard to the CBP material, temperature was recorded daily and moisture weekly in the feeding and resting areas during each experimental period. Temperature was measured at 15 cm depth with a thermometer and a 15-cm sounding line (K/JR-200 + 800 °C, Ventix, Sant Adrià de Besós, Spain). Sampling for chemical analysis was performed at week 11. Samples were collected at 15 cm depth, from the middle of the feeding and resting area of each pen, and later composited by pen. The amount of sample collected by pen was 5 kg. Samples were stored at −18 °C until analysis. Thus, we collected 4 samples per each material and period. Finally, the sampling of the piles was carried out every 10 days, collecting in total 9 samples of 6 kg each, and then samples were stored at −18 °C until analysis. Temperature was measured at 40 cm and 1 m depths with a thermometer and a 1-m sounding line (K/JR-200 + 800 °C, Ventix, Sant Adrià de Besós, Spain). Three samples were taken from each pile in the middle of 3 of the 4 sides of the truncated pyramid, at different depths to obtain a representative sample from each side, and were then composited by pile. The fourth side was next to a wall to facilitate the containment and prevention of landslide, making it inaccessible. In addition, during the composting process, the evolution of wet bulk density was monitored by weighing material of a known volume, according to [13]. Wet bulk density values were transformed to dry bulk density values by multiplying them by the corresponding dry matter (DM) content of samples.

### 2.3. Chemical Analysis

At the moment of analysis, samples were defrosted at room temperature. A watery extract was obtained from each fresh sample using 40 g and 200 mL of distilled water. After 30 min of stirring, the extract was centrifuged at 3600 rpm for 15 min. In the supernatant, the following determinations were carried out: pH and electric conductivity (EC) using a pH meter (Crison GLP 21; Hach Lange Spain, S.L.U., L'Hospitalet de Llobregat, Spain) and a conductivity meter (Crison GLP 31; Hach Lange Spain, S.L.U., L'Hospitalet de Llobregat, Spain), and ammonia nitrogen by means of an ammonium selective electrode (Orion 9512, Termo Fisher Scientific, Barcelona, Spain). Moisture content was determined by drying samples for 24 h at 103 °C. Ash content was determined in the dry and ground (1-mm screen) sample by loss on ignition at 600 °C for 2 h based on the AOAC method 942.05 [14] to ascertain the organic matter (OM) content of samples. Chemical stability degree (SD), as a way to predict composting ability, was determined by Klason lignin determination in accordance with [15]. In dry and ground samples (1-mm), nitrogen content was determined by the Kjeldahl procedure based on the AOAC method 976.05 [14]. The C/N ratio was estimated from the OM and nitrogen content in accordance with [16]. Mineral nutrient content was determined by flame photometry in the case of K (Model 410, Corning, Halstead, UK) and by spectrophotometry in the case of P (Model Cary 60, Agilent Technologies, Singapore, Malaysia) after dissolution of ash obtained from the ignition of samples at 470 °C in 3 N $HNO_3$.

### 2.4. Statistical Analysis

Chemical composition of raw bedding materials was compared using the GLM procedure of SAS (v. 9.3; SAS Institute Inc., Cary, NC, USA, 2011). Data related to the CBP material were analyzed by using the MIXED procedure of SAS (v. 9.3; SAS Institute Inc., Cary, NC, USA, 2011). The model contained the fixed effects of treatment, period and treatment × period interaction, and the random effect of pen was nested within the sequence, where the sequence is the order in which treatment is applied to the experimental unit. The Tukey multiple comparison test was applied to conduct mean separation across treatments and periods when the treatment × period interaction was significant. Regression analyses were performed to obtain the equations and the coefficients of determination between variables studied in the composted piles and the sampling days using the REG procedure of SAS (v. 9.3; SAS Institute Inc., Cary, NC, USA, 2011). The scopes of linear regression, obtained for each composted pile, were compared by means of a *t*-test after checking the homogeneity of variances.

## 3. Results and Discussion

### 3.1. Comparison between Bedding Materials

Sawdust obtained from sawmills is a bedding material commonly used in CBP barns for dairy cows [17,18], while FB is an alternative material that, in the present study, was composed of tree bark and vegetal fibers from a Mediterranean forest. Both materials showed an acidic pH, a low EC value, high OM content and low nutrient content, resulting in a high C/N ratio (Table 1). These characteristics are common of bedding materials used in CBP, together with a low moisture to assure the absorption capacity of animal urine and a particle size < 25 mm to promote the microbial activity due to the increased growing surface [11,18]. However, certain differences between both materials can be highlighted regarding their physical and chemical properties. Sawdust was slightly acidic with a higher EC value and OM content and lower in moisture and nutrients, resulting in a higher C/N ratio in S than in FB (Table 1; $p < 0.001$). In addition, S showed a lower moisture than FB and a finer granulometry because the proportion of particles < 2 mm was 49.2% and 22.0% for S and FB, respectively.

**Table 1.** Chemical composition and granulometry of raw bedding materials.

| Item | Material | | SEM | *p*-Value |
|---|---|---|---|---|
| | S [1] | FB [2] | | |
| pH | 5.01 | 6.01 | 0.112 | 0.001 |
| EC, mS/cm | 0.43 | 0.23 | 0.011 | 0.001 |
| Moisture, % | 10.2 | 32.1 | 1.97 | 0.001 |
| OM, % DM | 99.1 | 88.6 | 0.43 | 0.001 |
| Stability, % | 28.6 | 52.2 | 0.22 | 0.001 |
| Organic N, % DM | 0.20 | 0.35 | 0.026 | 0.001 |
| C/N ratio | 286 | 131 | 20.0 | 0.001 |
| P, % DM | 0.006 | 0.024 | 0.0020 | 0.001 |
| K, % DM | 0.054 | 0.131 | 0.0088 | 0.001 |
| Particle size, % | | | | |
| >25 mm | 0.0 | 0.0 | 0.02 | 0.434 |
| 25–12.5 mm | 0.6 | 8.6 | 0.76 | 0.001 |
| 12.5–10 mm | 0.7 | 6.2 | 0.50 | 0.001 |
| 10–6.3 mm | 8.3 | 13.2 | 0.74 | 0.001 |
| 6.3–5 mm | 6.8 | 10.3 | 1.26 | 0.077 |
| 5–2 mm | 34.3 | 39.7 | 2.99 | 0.232 |
| <2 mm | 49.2 | 22.0 | 4.62 | 0.002 |

[1] S = sawdust. [2] FB = forest biomass.

Stability was greater in FB than in S (Table 1; $p < 0.001$). The authors of [19] pointed out that the lignin content of hardwoods (angiosperms) is usually in the range of 18–25%, whereas this content ranges between 25 and 35% in the case of softwoods (gymnosperms). These higher values can be attributed to the great lignin content of pine bark, between 38 and 58%, an ingredient visually present in FB.

The content of C and N and the C/N ratio of the sawdust used in the present experiment was similar to that reported by [11], but with a P and K content ten times lower. In the case of FB, there is a lack of information with regard to this material when used in CBP barns. A similar bedding material obtained from the forest (conifer forest litter) to be used in CBP was described by [17], with similar C and macronutrient contents, and with a C/N ratio more suitable than sawdust for a composting process and greater than 50 to minimize ammonia losses [20]. However, the material described by [17] and the FB used in the present experiment differ in their granulometry. In our case, 61.7% of the material was less than 5 mm, while this percentage was lower for the conifer forest litter. This, together with the remaining particle fractions, suggests that the forest material described by [17] was more heterogenous and coarse than FB, probably due to the grinding machinery used. Differences in granulometry explain the differences in bulk density, which were 123 and 240 kg/m$^3$ for conifer forest litter and FB, respectively.

### 3.2. Comparison between Compost-Bedded Pack (CBP)

#### 3.2.1. Temperature and Moisture

Temperature and moisture affect the biological transformation of organic materials, particularly in aerobic conditions. Moisture and oxygen availability allow the microbial activity which leads to a temperature increase due to exogenous chemical reactions of molecule degradation. In a process such as composting, this temperature increase is observed as an indicator of an adequate evolution. However, in the case of CBP, the objective was not the biological transformation of the material as in the composting but the availability of safe housing for animal breeding. The presence of a C-rich material and the availability of a N source from feces and urine, moisture and oxygen provide the conditions for microbial growth and, consequently, a temperature increase.

Climatic conditions during the CBP process are shown in Table 2. The average ambient temperature was 14.7 ± 3.63 °C in Period 1 and decreased from 19.5 to 9.1 °C. In Period 2, the average ambient temperature was 11.2 ± 2.49 °C and increased from 5.9 to 14.1 °C Thus, Period 2 was colder than Period 1, as reflected in the lower temperatures recorded in CBP in Period 2. In Period 1, the mean CBP temperature was 34.7 ± 5.64 °C in CBP-S and 28.3 ± 6.87 °C in CBP-FB, and in Period 2, 31.2 ± 5.88 °C and 25.9 ± 3.08 °C in CBP-S and CBP-FB, respectively. The lower CBP temperatures recorded in CBP-FB would indicate a lower microbial activity due to the greater OM stability of this material and bigger particle size. However, in both cases, these temperatures were below the recommended range values for an effective composting [10,21,22], although this frequently occurs in the context of CBP management [7,11,23]. Average environmental humidity was 81.5 ± 4.21 in Period 1 and ranged between 74.1 and 90.4%. In Period 2, average environmental humidity was 74.5 ± 7.18% and ranged between 61.6 and 87.3%. In CBP-S, the initial moisture was 10% on average and the mean value was 54.1 ± 16.97% in Period 1, and 55.0 ± 16.43% in Period 2. In CBP-FB, the initial moisture was 32% on average and the mean moisture was 62.0 ± 14.00% in Period 1, and 59.4 ± 7.21% in Period 2. A similar moisture content was expected in both CBPs because we decided to add new bedding material when the moisture was greater than 60%, applying an average amount of 7.8 kg/d and 7.9 kg/d in CBP-S and CBP-FB, respectively. However, the moisture was slightly higher in CBP-FB than in CBP-S. The moisture and granulometry of FB could also have contributed to this final result.

**Table 2.** Climatic conditions, temperature and moisture content of compost-bedded pack (CBP) made with sawdust (CBP-S) and forest biomass (CBP-FB) during the time periods studied.

| | **Period 1** | **Period 2** |
|---|---|---|
| Climatic conditions | | |
| Temperature, °C | 14.7 ± 3.63 | 11.2 ± 2.49 |
| Humidity, % | 81.5 ± 4.21 | 74.5 ± 7.18 |
| CBP temperature, °C | | |
| CBP-S | 34.7 ± 5.64 | 31.2 ± 5.88 |
| CBP-FB | 28.3 ± 6.87 | 25.9 ± 3.08 |
| CBP moisture, % | | |
| CBP-S | 54.1 ± 16.97 | 55.0 ± 16.43 |
| CBP-FB | 62.0 ± 14.00 | 59.4 ± 7.21 |

#### 3.2.2. Chemical Characteristics

The chemical composition of CBP samples at week 11 is shown in Table 3. The pH values recorded were basic and there were increases in the pH and EC values from the raw bedding materials to CBP samples. This increase can be attributed to feces and urine provided by the cows. The pH and EC values were affected by the treatment × period interaction ($p$ = 0.001 and $p$ = 0.003, respectively). In Period 1, CBP-FB pH was higher than CBP-S pH, whereas CBP-FB EC was lower than CBP-S EC. In contrast, CBP-FB pH was lower than CBP-S pH in Period 2, whereas CBP-FB EC was not different from CBP-S

EC. The low EC value recorded in CBP-FB in Period 1 could be due to its higher moisture resulting in a higher dilution rate because this determination was made with a wet sample.

**Table 3.** Chemical composition of compost-bedded packs (CBP) after 11 weeks of use.

| Item | Period 1 | | Period 2 | | SEM | *p*-Value | | |
|---|---|---|---|---|---|---|---|---|
| | CBP-S [1] | CBP-FB [2] | CBP-S | CBP-FB | | T | P | T × P |
| pH | 8.00 [c] | 8.43 [b] | 8.96 [a] | 7.81 [c] | 0.078 | 0.001 | 0.021 | 0.001 |
| EC, mS/cm | 3.78 [a] | 2.39 [b] | 3.84 [a] | 3.79 [a] | 0.163 | 0.001 | 0.001 | 0.003 |
| Moisture, % | 61.3 [b] | 69.7 [a] | 63.7 [b] | 63.2 [b] | 1.09 | 0.002 | 0.037 | 0.001 |
| OM, % DM | 91.1 [a] | 83.4 [c] | 89.9 [a] | 85.7 [b] | 0.54 | 0.001 | 0.115 | 0.008 |
| Stability, % | ND | ND | 31.7 | 49.2 | 1.25 | 0.001 | ND | ND |
| N, % DM | | | | | | | | |
| Organic | 0.89 | 1.12 | 1.07 | 1.21 | 0.036 | 0.001 | 0.002 | 0.101 |
| Ammonia | 0.07 | 0.08 | 0.05 | 0.04 | 0.014 | 0.999 | 0.034 | 0.444 |
| Total | 0.96 | 1.19 | 1.12 | 1.24 | 0.039 | 0.001 | 0.005 | 0.109 |
| C/N ratio | 48:1 [a] | 35:1 [c] | 40:1 [b] | 35:1 [c] | 1.1 | 0.001 | 0.002 | 0.005 |
| P, % DM | 0.24 [b] | 0.28 [a] | 0.28 [a] | 0.25 [a,b] | 0.009 | 0.879 | 0.458 | 0.003 |
| K, % DM | 1.22 [b] | 1.64 [a] | 1.78 [a] | 1.62 [a] | 0.098 | 0.038 | 0.001 | 0.015 |

[1] CBP-S = compost-bedded pack made with sawdust. [2] CBP-FB = compost-bedded pack made with forest biomass. [a–c] Means within a row with different superscripts differ ($p < 0.05$).

The treatment × period interaction also affected the OM content ($p = 0.008$). The OM content was higher in CBP-S than in CBP-FB in both periods, but this difference was greater in Period 1 than in Period 2. The content of organic N was affected by treatment ($p = 0.001$), being higher in CBP-FB than in CBP-S. These results could be due to the differences detected in the raw bedding materials where OM was higher in S than in FB, and organic N was higher in FB than in S (Table 1). The period also affected organic N ($p = 0.002$), being higher in Period 2 than in Period 1. With regard to ammonia N, this content was affected by period ($p = 0.034$), being greater in Period 1 than in Period 2. However, these differences are not relevant from the agronomic point of view. In addition, there was a treatment × period interaction in the C/N ratio ($p = 0.005$). Although this ratio was greater in CBP-S than in CBP-FB in both periods ($p = 0.001$), in agreement with the differences detected in the raw bedding materials, and it was greater in Period 1 than in Period 2 ($p = 0.002$), the difference between treatments was greater in Period 1 than in Period 2.

The contents of the other macronutrients (P, K) were affected by the treatment × period interaction ($p = 0.003$ and $p = 0.015$, respectively). For P and K, the content of CBP-FB was greater than in CBP-S in Period 1, but the content between treatments was not different in Period 2. However, considering the higher differences found between raw bedding materials, where the P and K content was higher in FB than in S, the differences between CBPs were lower and in accordance with the raw materials in Period 1, or were not observed in Period 2. In any case, these differences are not relevant from the agronomic point of view.

The pH value, EC, moisture and the contents of N, P and K increased in CBP samples with regard to the raw bedding materials with high OM content, due to the high macronutrient content supplied by feces and urine. This would explain the C/N ratio decrease in CBP. Thus, the final characteristics of CBP will be linked with the bedding material added throughout the process. Taking into account the amount added (7.8 kg/d and 7.9 kg/d in CBP-S and CBP-FB, respectively) and the bulk density of the bedding material, the amounts added were 15.6 $m^3$/cow/year and 12.0 $m^3$/cow/year, for CBP-S and CBP-FB, respectively. These values are inside the range reviewed by [18]. In addition, the chemical characteristics found in the present experiment for CBP-S are in the range found by [11] and [24] for wood shavings and wood chips.

The SD of the OM in CBP, measured in Period 2, was greater in CBP-FB than in CBP-S, in agreement with the differences recorded in the raw bedding materials. The SD measured in CBP-S remained far away from 50, the threshold used to consider a compost mature. Taken together with the temperature evolution of CBP, this suggests that there was not a real composting process.

### 3.3. *Comparison between the Composted Piles*

#### 3.3.1. Temperature, Moisture and Bulk Density

From the beginning to the end of the composting period, ambient temperature increased from 22 to 26 °C and environmental humidity from 47 to 72%. The initial temperatures of the composting piles were 35.0 °C and 30.4 °C for S and FB, respectively, the final temperatures being 46.0 °C and 44.5 °C (Figure 1a). These temperatures were obtained after averaging data recorded at depths of 40 cm and 1 m. In both piles, a temperature increase was observed during the composting process, achieving the thermophile phase (>40 °C), and reaching sanitation temperature for S (≥55 °C more than 14 d according to European Regulation 2019/1009), while FB remained below this limit. The temperature achieved at d 90 in both piles would indicate that the composting process had not ended, probably due to the low C/N ratio and the lignocellulosic content of both materials, particularly in FB. The initial moisture of the S pile and the FB pile was 61.5% and 62.2%, respectively, whereas the final moisture was 47.9% and 55.4% (Figure 1b). The values over the period were adequate for a composting process, being over 50% in order to maintain microbial activity [25]. The piles were watered periodically, which helped to conserve the moisture. The bulk density evolution is shown in Figure 2. Dry bulk density increased slightly, and changes were due to mineralization and particle size reduction. Bulk density was higher in the FB pile than in the S pile.

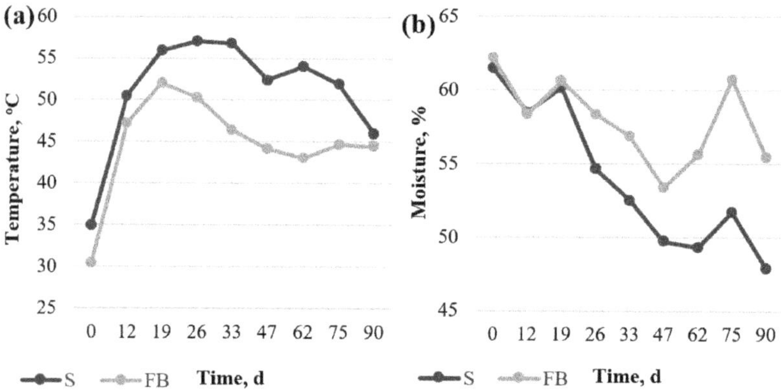


**Figure 1.** Effect of treatment (sawdust = S and, forest biomass = FB) on temperature (**a**) and moisture (**b**) of the composted piles.

#### 3.3.2. Chemical Characteristics

Initial and final values of the variables measured in the piles are shown in Table 4. In a composting process, an initial acidification followed by a gradual alkalinization is a common evolution of the pH medium. However, in the present experiment, the changes in the pH values of the piles were small and did not respond to this pattern (Figure 3). The pH values were more basic in S than in FB, which implies greater risk of N losses in S. With regard to EC, an increase in its value was expected during the composting process due to OM mineralization. This happened in S but not in FB, proving that composting was less strong in the FB pile.

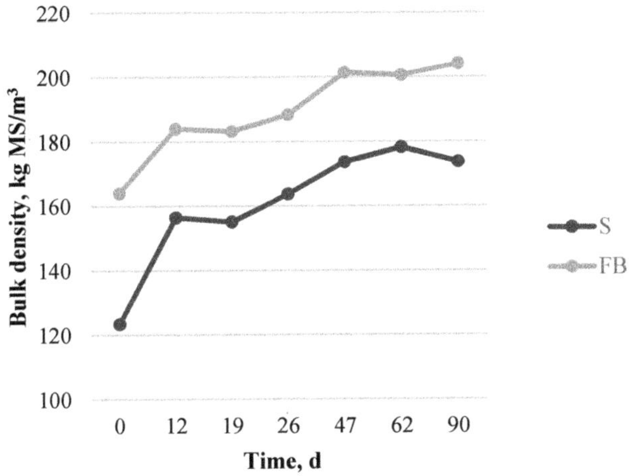


**Figure 2.** Bulk density evolution of samples taken from the composted piles (sawdust = S and forest biomass = FB).

**Table 4.** Initial and final values (mean ± SD) of the variables measured in the composted piles made with sawdust (S) and forest biomass (FB).

| Item | S | | FB | |
|---|---|---|---|---|
| | Day 0 | Day 90 | Day 0 | Day 90 |
| pH | 8.72 ± 0.017 | 8.39 ± 0.035 | 7.45 ± 0.015 | 7.72 ± 0.100 |
| EC, mS/cm | 3.59 ± 0.145 | 5.58 ± 0.106 | 3.39 ± 0.025 | 3.10 ± 0.061 |
| Moisture, % | 61.5 ± 0.24 | 47.9 ± 0.15 | 62.2 ± 0.29 | 55.4 ± 0.18 |
| OM, % DM | 91.7 ± 0.18 | 87.4 ± 0.13 | 86.3 ± 0.30 | 82.9 ± 0.03 |
| Stability, % | 31.7 | 34.9 | 49.2 | 53.9 |
| N, % DM | | | | |
| Organic | 0.88 ± 0.039 | 1.05 ± 0.058 | 1.16 ± 0.013 | 1.27 ± 0.054 |
| Ammonia | 0.049 ± 0.0004 | 0.004 ± 0.0002 | 0.013 ± 0.0006 | 0.001 ± 0.0005 |
| C/N ratio | 52.0 | 41.8 | 37.2 | 32.6 |
| P, % DM | 0.24 ± 0.017 | 0.31 ± 0.040 | 0.27 ± 0.021 | 0.24 ± 0.011 |
| K, % DM | 1.74 ± 0.071 | 2.02 ± 0.134 | 1.94 ± 0.269 | 1.87 ± 0.066 |

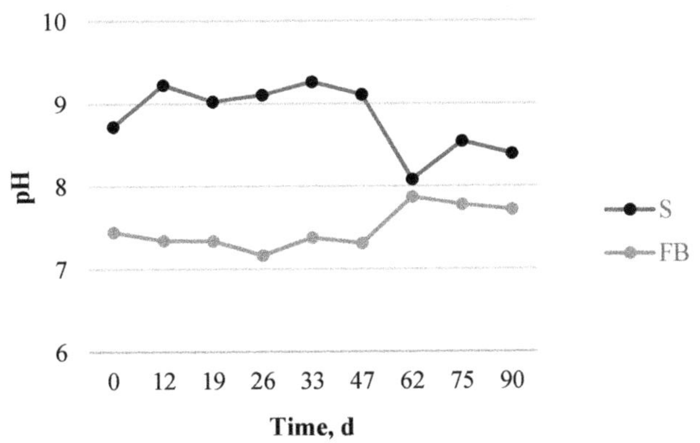


**Figure 3.** Evolution of pH of samples taken from the composted piles (sawdust = S and forest biomass = FB).

Although a small decrease in the OM in both piles was observed, the content was high in both cases at d 90, which would indicate only a partial transformation of this OM during composting. This could be explained by the high C/N ratio recorded at d 0, which reflected a N lack, and by the high presence of lignin in wood waste, which would explain the difficulty in the OM degradation. Thus, at the end of the process, a large quantity of OM remained non-degraded. Nevertheless, a slight increase in stability was observed, from 31.7 to 34.9% and 49.2 to 53.9%, respectively, for S and FB. According to [15], the value over 50% recorded in the FB pile at d 90 would indicate that the material was chemically stable and would produce a progressive degradation of OM in an eventual soil application and a consequent gradual release of nutrients. There was a slight increase in the content of N, P and K from d 0 to d 90 in the S pile, which can be explained by a concentration process due to the OM decrease. In the FB pile, this slight increase was only observed in the N content. The initial ammonia N content was low in both piles, in agreement with their low total N content. Although it was lower in FB than in S, its evolution over time was similar to that normally observed in a composting process because after a slight increase during the first days, it then decreased till d 90 (Figure 4). This decrease could be explained by utilization by microorganisms, the nitrification process or the loss caused by turning [26]. All these values were close to or below the upper limit proposed by [16] to consider a compost being mature.

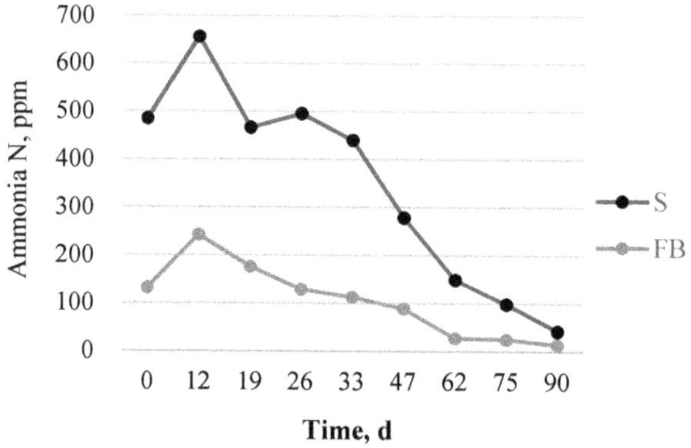


**Figure 4.** Evolution of ammonia N content of samples taken from the composted piles (sawdust = S and forest biomass = FB).

In order to accelerate the composting process, it would be necessary to increase the N content in the original material by adding a N source to reduce its initial C/N ratio. In this sense, the practice used by some farmers of drying the material resulting in the CBP and reusing it in the barn could help achieve this objective. However, effort should be made to achieve the appropriate conditions to ensure the sanitation of material and avoid animal health problems. Another alternative to achieve this sanitation, when lignocellulosic materials are used, could be to lengthen the process (>6 months).

The progression of each chemical compound of these piles over time, expressed with the corresponding regression equation, is shown in Table 5. The relationship was statistically significant in both piles for OM, organic N, ammonia N and the C/N ratio, the slope being negative for OM, ammonia N and the C/N ratio, and positive for organic N. In those variables, the slope comparison showed that there were no statistical differences between treatments (data not shown). Moreover, in the sawdust pile, the relationship exists also for EC, moisture and total N, and in the forest biomass pile, for pH.

**Table 5.** Regression equations in each composting pile (sawdust = S; forest biomass = FB) between chemical variables and time (d) of composting period.

| Item | S | | | | FB | | | |
|---|---|---|---|---|---|---|---|---|
| | Equation | $R^2$ | *p*-Value | RMSE | Equation | $R^2$ | *p*-Value | RMSE |
| pH | 9.19 − 0.009d | 0.41 | 0.064 | 0.343 | 7.25 + 0.006d | 0.53 | 0.026 | 0.178 |
| EC, mS/cm | 3.66 + 0.023d | 0.95 | 0.001 | 0.176 | 2.03 + 0.001d | 0.02 | 0.745 | 0.252 |
| Moisture,% | 59.9 − 0.146d | 0.79 | 0.002 | 2.468 | 59.9 − 0.047d | 0.25 | 0.174 | 2.681 |
| OM, % DM | 91.0 − 0.041d | 0.90 | 0.001 | 0.499 | 85.3 − 0.027d | 0.65 | 0.028 | 0.688 |
| N, % DM | | | | | | | | |
| Organic | 0.88 + 0.003d | 0.68 | 0.006 | 0.061 | 1.14 + 0.001d | 0.48 | 0.039 | 0.037 |
| Ammonia | 0.06 − 0.001d | 0.89 | 0.001 | 0.008 | 0.02 − 0.001d | 0.75 | 0.002 | 0.004 |
| C/N | 52.0 − 0.145d | 0.84 | 0.004 | 2.286 | 37.5 − 0.050d | 0.73 | 0.015 | 1.073 |
| P, % DM | 0.26 + 0.001d | 0.63 | 0.107 | 0.019 | 0.28 − 0.001d | 0.27 | 0.292 | 0.028 |
| K, % DM | 1.83 + 0.004d | 0.71 | 0.073 | 0.107 | 1.76 + 0.002d | 0.29 | 0.270 | 0.127 |

### 3.4. Agronomic Suitability of the Materials

Both CBP materials contain high amounts of OM and a low nutrient content in comparison with cattle manure. Thus, they must be considered organic fertilizers to provide OM to the soil. The CBP review by [18] regards this material as a green waste compost, and stated that in the long term, the use of CBP as manure can result in considerably higher amounts of OM and a larger accumulation of N than cattle manure. The C/N ratio of both CBPs was high but within the wide range, from 10.5 to 49.3, reviewed by [18]. In addition, high ratios are expected when wood-derived materials are considered [11]. The incorporation in the soil of materials with a high C/N ratio can initially lead to a N inorganic immobilization because it is used by microbes, making it temporarily inaccessible to crops. This negative effect could happen if the material of the S pile was used due to its less stable OM, but not in the case of the FB pile material with a more stable OM. However, in accordance with the Spanish regulation RD506/2013 on fertilizing, both materials would be unmarketable because of a C/N ratio greater than 20. Thus, this final composting process did not bring about any meaningful change to CBP materials. Only if sanitation was achieved could this additional work be justified to add safety in cattle farms.

## 4. Conclusions

From the agronomic point of view, although both materials had a high C/N ratio to be registered as soil fertilizers, they present potentially valuable characteristics when it comes to organically amending the soil, as they are high in organic matter content and low in nutrients. Further, the organic matter of the forest biomass pile was well stabilized (more than 50%). Therefore, as sawdust and forest biomass are very likely to improve soil properties when added to the land, further studies focusing on improving the composting management should be made to achieve effective composting, assuring an adequate C/N ratio and the sanitation of this material in the case of its reutilization as a compost-bedded pack.

**Author Contributions:** Conceptualization, L.C. and A.F.; formal analysis, L.L. and C.G.; investigation, L.L., C.G., M.L. and T.B.; resources, M.L., L.C., A.F. and T.B.; data curation, L.L., C.G. and M.L.; writing—original draft preparation, L.L., C.G. and T.B.; writing—review and editing, L.L., L.C., A.F. and T.B.; visualization, L.L., A.F. and T.B.; supervision, M.L., L.C., A.F. and T.B.; project administration, L.C. and A.F.; funding acquisition, L.C. and A.F. All authors have read and agreed to the published version of the manuscript.

**Funding:** This research was funded by the Spanish Ministry of Economy and Competitiveness and the European Regional Development Fund, reference project AGL2015-68373-C2-1-R, Madrid, Spain.

**Institutional Review Board Statement:** Animal procedures were approved by the Institutional Animal Care and Use Committee (reference CEEAH 9963) of the Universitat Autònoma de Barcelona (Spain) in accordance with the European directive 2010/63/EU.

**Informed Consent Statement:** Not applicable.

**Data Availability Statement:** The data presented in this study are available on request from the corresponding author.

**Acknowledgments:** The authors would like to thank the UAB Farms and Experimental Fields Service for supporting the experimental work activities.

**Conflicts of Interest:** The authors declare no conflict of interest.

## References

1. 68% of the World Population Projected to Live in Urban Areas by 2050. Available online: https://www.un.org/development/desa/en/news/population/2018-revision-of-world-urbanization-prospects.html (accessed on 18 February 2021).
2. Boogaard, B.K.; Oosting, S.J.; Bock, B.B.; Wiskerke, J.S.C. The sociocultural sustainability of livestock farming: An inquiry into social perceptions of dairy farming. *Animal* **2011**, *5*, 1458–1466. [CrossRef] [PubMed]
3. Song, X.P.; Hansen, M.C.; Stehman, S.V.; Potapov, P.V.; Tyukavina, A.; Vermote, E.F.; Townshend, J.R. Global land change from 1982 to 2016. *Nature* **2018**, *560*, 639–643. [CrossRef] [PubMed]
4. Minnich, R.A. An integrated model of two fire regimes. *Conserv. Biol.* **2001**, *15*, 1549–1553. [CrossRef]
5. Oliveras, I.; Piñol, J.; Viegas, D.X. Modelling the long term effect of changes in fire frequency on the total area burnt. *Orsis* **2005**, *20*, 73–81. Available online: https://www.raco.cat/index.php/Orsis/article/view/51789 (accessed on 18 February 2021).
6. Barberg, A.E.; Endres, M.I.; Salfer, J.A.; Reneau, J.K. Performance and welfare of dairy cows in an alternative housing system in Minnesota. *J. Dairy Sci.* **2007**, *90*, 1575–1583. [CrossRef]
7. Black, R.A.; Taraba, J.L.; Day, G.B.; Damasceno, F.A.; Bewley, J.M. Compost bedded pack dairy barn management, performance, and producer satisfaction. *J. Dairy Sci.* **2013**, *96*, 8060–8074. [CrossRef] [PubMed]
8. Fernández, A.; Mainau, E.; Manteca, X.; Siurana, A.; Castillejos, L. Impacts of compost bedded pack barns on the welfare and comfort of dairy cows. *Animals* **2020**, *10*, 431. [CrossRef] [PubMed]
9. Barberg, A.E.; Endres, M.I.; Janni, K.A. Compost dairy barns in Minnesota: A descriptive study. *Appl. Eng. Agric.* **2007**, *23*, 231–238. [CrossRef]
10. Janni, K.A.; Endres, M.I.; Reneau, J.K.; Schoper, W.W. Compost dairy barn layout and management recommendations. *Appl. Eng. Agric.* **2007**, *23*, 97–102. [CrossRef]
11. Shane, E.M.; Endres, M.I.; Janni, K.A. Alternative bedding materials for compost bedded pack barns in Minnesota: A descriptive study. *Appl. Eng. Agric.* **2010**, *26*, 465–473. [CrossRef]
12. Llonch, L.; Castillejos, L.; Mainau, E.; Manteca, X.; Ferret, A. Effect of forest biomass as bedding material on compost-bedded pack performance, microbial content, and behavior of nonlactating dairy cows. *J. Dairy Sci.* **2020**, *103*, 10676–10688. [CrossRef]
13. Huerta-Pujol, O.; Soliva, M.; Martínez-Farré, F.X.; Valero, J.; López, M. Bulk density determination as a simple and complementary tool in composting process control. *Bioresour. Technol.* **2010**, *101*, 995–1001. [CrossRef]
14. AOAC. *Official Methods of Analysis*, 15th ed.; Association of Official Analytical Chemists: Washington, DC, USA, 1990.
15. López, M.; Huerta-Pujol, O.; Martínez-Farré, F.X.; Soliva, M. Approaching compost stability from Klason lignin modified method: Chemical stability degree for OM and N quality assessment. *Resour. Conserv. Recycl.* **2010**, *55*, 171–181. [CrossRef]
16. Zucconi, F.; de Bertoldi, M. Compost specifications for the production and characterization of compost from municipal solid waste. In *Compost: Production, Quality and Use*; de Bertoldi, M., Ferranti, M.P., l'Hermite, P., Zucconi, F., Eds.; Elsevier: Barking, UK, 1987; pp. 30–50.
17. Ferraz, P.F.P.; Ferraz, G.A.S.; Leso, L.; Klopcic, M.; Barbari, M.; Rossi, G. Properties of conventional and alternative bedding materials for dairy cattle. *J. Dairy Sci.* **2020**, *103*, 8661–8674. [CrossRef]
18. Leso, L.; Barbari, M.; Lopes, M.A.; Damasceno, F.A.; Galama, P.; Taraba, J.L.; Kuipers, A. Invited review: Compost-bedded pack barns for dairy cows. *J. Dairy Sci.* **2020**, *103*, 1072–1099. [CrossRef] [PubMed]
19. Rowell, R.M.; Pettersen, R.; Tshabalala, M.A. Cell Wall Chemistry. In *Handbook of Wood Chemistry and Wood Composites*, 2nd ed.; Rowell, R.M., Ed.; CRC Press: Boca Raton, FL, USA, 2012.
20. Kirchman, H. Losses, plant uptake and utilisation of manure nitrogen during a production cycle. *Acta Agric. Scand.* **1985**, *24*, 1–77.
21. Rynk, R.; van de Kamp, M.; Willson, G.B.; Singley, M.E.; Richard, T.L.; Kolega, J.J.; Gouin, F.R.; Laliberty, L.; Kay, D.; Murphy, D.W.; et al. *On-Farm Composting Handbook (NRAES 54)*; Northeast Regional Agricultural Engineering Service (NRAES): New York, NY, USA, 1992.
22. Guidelines for Managing Compost Bedded-Pack Barns. Available online: https://www.dairypc.org/catalog/guidelines-for-managing-compost-bedded-pack-barns (accessed on 4 June 2014).
23. Leso, L.; Uberti, M.; Morshed, W.; Barbari, M. A survey on italian compost dairy barns. *J. Agric. Eng.* **2013**, *44*, 203–207. [CrossRef]
24. Eckelkamp, E.A.; Taraba, J.L.; Akers, K.A.; Harmon, R.J.; Bewley, J.M. Understanding compost bedded pack barns: Interactions among environmental factors, bedding characteristics, and udder health. *Livest. Sci.* **2016**, *190*, 35–42. [CrossRef]

25. Haug, R.T. *The Practical Handbook of Compost Engineering*, 1st ed.; CRC Press: Boca Raton, FL, USA, 1993.
26. DeLaune, P.B.; Moore, P.A.; Daniel, T.C.; Lemunyon, J.L. Effect of chemical and microbial amendments on ammonia volatilization from composting poultry litter. *J. Environ. Qual.* **2004**, *33*, 728–734. [CrossRef] [PubMed]

processes

MDPI

*Review*

# Gaseous Emissions from the Composting Process: Controlling Parameters and Strategies of Mitigation

**Tahseen Sayara [1] and Antoni Sánchez [2,*]**

[1] Department of Environment and Sustainable Agriculture, Faculty of Agricultural Sciences and Technology, Palestine Technical University-Kadoorie, Tulkarm 7, Palestine; tsayara@yahoo.com
[2] Departament d'Enginyeria Química, Biològica i Ambiental, Escola d'Enginyeria, Universitat Autònoma de Barcelona, Bellaterra, 08193 Barcelona, Spain
* Correspondence: antoni.sanchez@uab.cat

**Abstract:** Organic waste generation, collection, and management have become a crucial problem in modern and developing societies. Among the technologies proposed in a circular economy and sustainability framework, composting has reached a strong relevance in terms of clean technology that permits reintroducing organic matter to the systems. However, composting has also negative environmental impacts, some of them of social concern. This is the case of composting atmospheric emissions, especially in the case of greenhouse gases (GHG) and certain families of volatile organic compounds (VOC). They should be taken into account in any environmental assessment of composting as organic waste management technology. This review presents the relationship between composting operation and composting gaseous emissions, in addition to typical emission values for the main organic wastes that are being composted. Some novel mitigation technologies to reduce gaseous emissions from composting are also presented (use of biochar), although it is evident that a unique solution does not exist, given the variability of exhaust gases from composting.

**Keywords:** organic wastes; composting; gaseous emissions; mitigation strategies

**Citation:** Sayara, T.; Sánchez, A. Gaseous Emissions from the Composting Process: Controlling Parameters and Strategies of Mitigation. *Processes* **2021**, *9*, 1844. https://doi.org/10.3390/pr9101844

Academic Editor: Guining Lu

Received: 17 September 2021
Accepted: 14 October 2021
Published: 18 October 2021



## 1. Introduction

As a result of increasing solid wastes' generation, the implementation of a reliable technology to deal with these wastes is considered as a pillar of sustainable development of any nation [1]. However, the selection of any technology should be compatible with the economic situation within the jurisdiction. Concurrently, the used technology has to satisfy the laws and regulations that fundamentally aim to reduce any environmental and health problems [2]. Among the different technologies used in this field is the composting process, which has been used to deal with solid wastes and mainly for the organic fraction of wastes [3,4]. This process is recognized as an environmentally friendly and cost-effective method, as organic matter is biologically degraded under aerobic conditions [5]. This biodegradation of organic matter contributes to reducing the volume of wastes and producing a stabilized and nutrient-rich final end product, "compost", that could be used in agricultural activities due to its various positive impacts on the physical and chemical properties of the soil, meanwhile reducing utilization of inorganic fertilizers [6–8]. Actually, when the process-controlling parameters are well adjusted, this will lead to different advantages; thereby the process is viewed as a sustainable alternative for landfilling and other treatment options [9]. However, even though composting is a natural biochemical decomposition process, a successful composting operation that produces a valuable end product is normally associated with releasing gaseous emissions including greenhouse gases (GHGs) into the atmosphere (Figure 1). The released GHGs are attributed to energy requirements for composting plants' operation and to the biochemical reactions within the organic waste itself, which produces $CO_2$, methane ($CH_4$), and nitrous oxide ($N_2O$) due to the mineralization and degradation of organic matters [10,11]. According to Hao et al. [12], the majority of organic carbon is converted to $CO_2$, whereas the methane accounts for less than 6%. Nevertheless, it should be noted that even though $CO_2$ represents the major part of the emissions, it does not add to global warming due to the biogenic origin of carbon. On the contrary, the other emissions resulting from the process such as $CH_4$ and $N_2O$ have a direct impact on the global warming, while $NH_3$, Sulphur compounds, and most of the volatile organic compounds (VOCs) emissions cause undesirable and other odor nuisances [9,13,14]. Indeed, these gases contribute to climate change, global warming, acidification, and eutrophication of ecosystems as a result of $NH_3$ deposition, which also contributes to the formation of particulate matters in the air [9]. As a matter of fact, these GHGs and ammonia ($NH_3$) deteriorate the compost quality, besides being a secondary environmental pollution, as mentioned before [15,16].

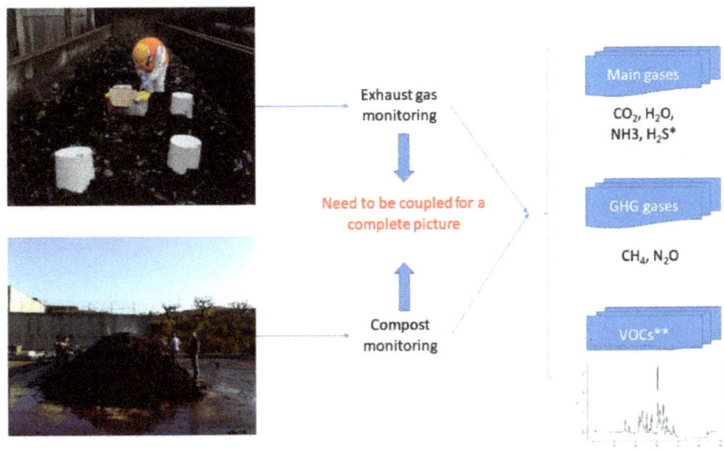


**Figure 1.** Monitoring exhaust gases from a composting process. * $H_2S$ is only significantly observed when anaerobic conditions prevail in the composting process. ** VOCs: Volatile Organic Compounds, a wide group including families such as alcohols, aldehydes, alkanes, aromatic hydrocarbons, carboxylic acids, ketones, nitrogen compounds, phenols, sulphur compounds, and terpenes, among others.

Emissions are formed due to inadequate aerobic conditions of composting [9]. Generally, the creation of anaerobic zones in compost mixtures results in $CH_4$ emissions, whereas nitrogen transformation and loss ($NH_3$ and $N_2O$) are linked to ammonification, nitrification, and denitrification during the composting process [17–19], but still they are less than GHGs' emissions generated from landfilling and waste-to-energy processes [20–22]. During the composting process, various forms of VOCs are formed, where the rates and specific forms of these emissions highly depend on the feedstock materials and composting phases, taking into account that aeration of the composting mixture has a major role in releasing of these compounds [23,24]. The rate of gaseous emissions is differing based on the applied composting method, whereas the initial content of the carbon and nitrogen is of great importance in the produced amount of gaseous emissions such that low total carbon (C) and nitrogen (N) content can simultaneously reduce $CH_4$, $CO_2$, and $N_2O$ emissions [25]. In this regard, it has been reported that manure composting may account for 46% and 67% of the initial N and C content of the original manure, respectively [26]. The losses in nitrogen mass are normally in the form of $NH_3$ emissions, whereas nitrous oxide emission, which has 265 times the global warming potential of $CO_2$ [27], accounts for about 0.1–5% of total N losses [28–30]. Noteworthy is the amount of these emissions influenced by the composting technology. In this context, silo composting reduced GHGs' losses by

82.84% compared with turning composting, which resulted in larger carbon and nitrogen losses [25].

Reducing the impact of the resulting emissions from composting has been investigated and practiced using different approaches. For example, biofilters effectively reduced the $NH_3$ emission with mitigation efficiency (ME) of 97%, whereas adding sawdust or straw reduced $CH_4$ and $N_2O$ emissions by 66.3% and 44.0%, respectively, as such types of materials enhance the absorption and microbial assimilation of $NH_4^+/NH_3$ [25]. Providing an optimal initial mixture and maintaining aerobic conditions among other practices have been used to mitigated both odors and GHGs [31]. This research presents a comprehensive overview on the gaseous emissions from composting processes. Factors that influence the production of emissions and the mitigation approaches are highlighted also.

## 2. Gas Emissions from Composting Process

As a result of microbial activities and putrefaction, gaseous emissions from organic wastes are produced [10]. These emissions, which include $CO_2$, $CH_4$, $N_2O$, Sulphur compounds, and many other volatile organic compounds (VOCs), as shown in Table 1, have been detected during the different phases of the waste management [9,31].

**Table 1.** Volatile organic compounds (VOCs) detected in the composting of different organic wastes.

| Waste | Main VOC Family | Other VOCs | Reference |
|---|---|---|---|
| Poultry litter | Alkanes and alkylated benzenes | Aldehydes, terpenes, and ketones | [32] |
| Chicken manure and biochar | Ketones, phenols, and organic acids | Aliphatic, aromatic, and terpenes | [33] |
| Municipal solid waste | Alkylated benzenes, alcohols, and alkanes | - | [14] |
| Wastewater sludge | Terpenes | Furans and esters | [34] |
| Digested wastewater sludge | Terpenes | Alcohols and Ketones | [34] |
| Swine carcass | Sulphur compounds | - | [35] |
| Municipal solid waste | Terpenes | Alcohols, volatile fatty acids, and aromatic compounds | [36] |
| Livestock and Poultry Manure | Sulfur compounds, aliphatic hydrocarbons, aromatic hydrocarbons | Chlorinated organic compounds | [37] |
| Municipal solid waste digestate | Terpenes and oxygenated compounds | Sulphur compounds and methanethiol | [38] |
| Green waste | Alcohols | Alkenes, aliphatic alkanes, aromatic hydrocarbons, ketones, aldehydes, furans, and esters | [39] |
| Sewage sludge | Isovaleraldehyde, butyric acid, sulphur compounds, and pinene | Indole, skatole, and phenol | [40] |

In the composting process, the amount of emitted gases is highly influenced by the type of treated wastes, composting technology, and operational conditions, mainly aeration, which would have a direct impact in reducing the rate of emissions, mainly $N_2O$ and methane, when it is properly adjusted [9]. According to Goldstein [41], the odors generated from composting plants are attributed to different compounds including terpenes, alcohols, aldehydes, fatty acids, ammonia, and a range of Sulphur compounds. Methane is normally formed during the composting process due to anaerobic condition that could be established in some parts of the composted material such as middle zones of a pile, which suffer from insufficient diffusion of oxygen [12,24]. However, nitrous oxides are produced due to nitrification and denitrification [42], taking into account that other conditions such as temperature, nitrate concentration, and aerobic conditions influence these emissions [43].

For the determination of emissions' rates and their subsequent global impact, emission factor is usually used as a useful tool for VOCs, $NH_3$, or GHGs. The emission factor is usually expressed per ton of treated waste or per amount of obtained compost [44]. For instance, GHGs' emission factor, in terms of kg $CO_{2eq}.Mg^{-1}$ dry matter of sewage sludge (DM–SS), was found to be $2.30 \times 10^2$. On the other hand, the sewage sludge composting odor emission factor (OEF) was $2.68 \times 10^7$ ou.$Mg^{-1}$ DM–SS [45]. Different emission factors could be found in the literature, depending on the characteristics of the feedstock or the composting technology [46–49].

Compared with other treatments or management technologies for solid waste, different studies demonstrated that the composting process has less impact on global warming, as it produces lower amounts of GHGs. In this regard, Lou and Nair [50] showed that composting of municipal solid waste produced about 1.29 t $CO_2$-eq/t-of-waste, which is lower than the amount produced from landfills. Actually, this was concluded and documented by different studies, which emphasized that composting produces lower amounts of emissions (g $CO_2$-eq/t-of-waste) compared to landfilling and incineration based on emission factor [21,49–51]. However, when composting and vermicomposting were compared, it was found that the vermicomposting process caused 78.19% lesser GHGs' emission as compared to the composting process, which released 80.9 kg $CO_2$-eq/t-of-waste [52].

## 3. Factors Affecting the Emissions' Rates

During the initial stages of the composting process, both nitrogen and sulfur are in the organic form [53]. As the process proceeds forward, the mineralization of the organic nitrogen leads to the formation of ammonia ($NH_3$), which could react with hydrogen ions to form ammonium ($NH_4^+$). The $NH_4^+$-to-$NH_3$ equilibrium is highly affected by the dominant conditions within the composting mixture, mainly the pH value and temperature [54–56]. Thermophilic temperatures and alkaline conditions enhance the loss of nitrogen as ammonia. Additionally, ammonia-oxidizing bacteria or archaea and nitrite-oxidizing bacteria convert part of the nitrogen to nitrate through the nitrification process. This nitrate is used by the microbial community, but it would be converted to $N_2O$ under certain conditions including denitrifications' process, especially under insufficient oxygen levels [54]. Furthermore, the low levels of oxygen lead to the formation of some anaerobic zones within the compositing mixture. These zones play a major role in the sulfur transformation and the production of $H_2S$ through the action of Sulfate-reducing bacteria (anaerobic) during the degradation of the organic matter [57]. Additionally, during the formation of $H_2S$, other reduced sulfur compounds will also be produced, such as MeSH, $Me_2S$, $Me_2SS$, and others [54]. The following are some of the main factors that affect the emissions' rates during the composting process.

### 3.1. Composting Method

Composting of solid wastes can be carried out using different technologies [4]. As reported in various studies, the used technology has a direct impact on the rate of gaseous emissions [31]. In this regard, turned and windrow technologies showed higher values of $CH_4$, $CO_2$, and $N_2O$ compared with other technologies [25,58–60]. Turning of composted materials increases the chances of releasing trapped gasses within the composting materials and exposing them to the air. The frequent turning helps in re-structuring the materials and improving the porosity; thus, more air could be diffused that supports the microbial activities and promotes the biodegradation of the organic matter, which ultimately increases the amount of $CO_2$ volatilization [25,61,62]. Additionally, $N_2O$ emissions are high in turned piles compared with other technologies. This is attributed to the losses as a result of nitrification near the surface and denitrification by mixing $NO_3$ /$NO_2$ accumulated on the surface into the pile [63–65]. Amlinger et al. [44] suggested that high aeration and effective stripping of $NH_3$ during the early stages of composting can reduce $N_2O$ formation. Additionally, the enzymatic activity is thought to be affected by turning and increases the $N_2O$ emissions. Under anoxic conditions, denitrification enzymes are in an equilibrium state; however, when the material is exposed to oxygen as a result of turning, nitrous oxide reductases that can catalyze the transformation of $N_2O$ to $N_2$ are clearly more severely inhibited by $O_2$ than the other reductases, resulting in a stronger $N_2O$ emission [66,67]. Between turned piles and static systems, lower emissions of $N_2O$ and $CH_4$ were observed in turned piles, attributed to the difference to the anaerobic zones in the static system [68–70]. In silo composting, the $N_2O$ and $CH_4$ emissions were lower than other methods. This is because a good aeration system resulted in reducing the chances for the denitrification process. Moreover, more $NH_3$ is emitted from this system, which reduced the substrates for $N_2O$ emissions [25,59,60]. Similar observations regarding the aeration effect on the emissions were noticed by Ermolaev et al. [71], such that lower emissions of $CH_4$ and $NO_2$ were observed regardless of the amount of aeration.

### 3.2. Average Composting Temperature

The temperature evolution during the composting process has a direct impact on the rate of gaseous emissions. It is well documented that a positive correlation is normally observed between temperature and emissions' rate, where a higher rate of emissions was recorded with high temperatures and, more specifically, in the thermophilic ranges (45–70 °C) [9]. This could be attributed to the high rate of organic matter decomposition at higher temperature [25,62,72]. In this regard, Fillingham et al. [59] demonstrated that the highest $NH_3$ emissions were recorded in silo composting (111.07 g [$NH_3$-N] $kg^{-1}$ [TN]) compared to windrow composting, and the difference was attributed to high temperatures, which were about 65 °C in silo composting. These conditions enhance the equilibrium between ammonium and $NH_3$ towards gaseous $NH_3$, whereas low temperature inhibits microbial ammonization, thus reducing $NH_3$ emissions [25,63]. Similarly, elevated temperatures result in increasing $CH_4$ emissions. This could be explained by the high rate of microbial activities that result in increasing the temperatures, and these conditions are associated with high oxygen consumption, which ultimately leads to forming anaerobic conditions and the formation of $CH_4$ [45,72–74]. The same trend was also observed regarding $CO_2$ and $N_2O$ emissions that exhibit an increase with increasing temperature [63,75]. In this context, the maximum concentrations measured during sewage sludge composting were 2600 ppmv of $NH_3$, 66 ppmv of $H_2S$, and 1650 ppmv of tVOCs, which were observed during the peak of maximum temperature of the reactor [76]. Accordingly, controlling this parameter would help in controlling the emissions' rates [9].

### 3.3. Initial Moisture Content

Providing an optimum moisture content is crucial for the composting process performance, as it will promote the microbial activities [4]. However, increasing moisture content above the recommended values (40–60%) would result in creating anaerobic zones within the composted materials [25]. This was clear regarding $CH_4$ emissions that were positively correlated to the moisture content of the compost [28,77], meanwhile a negative correlation with moisture content was recorded regarding the $CO_2$ emissions [25]. However, for $N_2O$ emissions, the correlation with moisture content is not well established. For instance, Hwang and Hanaki [78] demonstrated that $N_2O$ emissions decreased when the material became very moist because of the inhibition of $N_2O$ nitrification, but Yan et al. [79] showed that the $N_2O$ would increase with water content as aerobic and anaerobic zones would simultaneously exist and also the nitrification and denitrification might be promoted concurrently and $N_2O$ emission flux could become relatively high.

### 3.4. Initial Total Carbon (TC) and Initial Total Nitrogen (TN) Content

Carbon and nitrogen are essential for the microbial activities during the composting process. Providing an adequate ratio of carbon and nitrogen (normally indicated as C/N ratio, with recommended values between 25:1 to 30:1) is considered as one of the controlling parameters in this process [25,31]. Importantly, these elements also have an impact on the rate of emissions resulting from the process. When the microbial communities biodegrade the organic matter under aerobic condition, most of the carbon is lost as $CO_2$, such that a linear relation between carbon content and $CO_2$ emissions would be observed during the process [76]. Furthermore, initial carbon content was found to have a positive correlation with $CH_4$ emissions [80], considering the nitrogen content, which is a primary source for methanogenic bacteria [11,81]. Similarly, the rate of $N_2O$ emissions is positively correlated with nitrogen content, as both nitrification and denitrification are enhanced by a high content of nitrogen [65,82]. Usually, composting feedstocks with low C/N ratios and high moisture contents provides favorable conditions of producing more greenhouse gas emissions [31]. Ammonia emissions are also affected by the C/N ratio [31,73,83,84].

### 3.5. Aeration Rate

As an aerobic process, supplying a sufficient amount of oxygen is recognized as an important parameter for maintaining the microbial activity and reducing the gas emission during the composting process [85]. Sufficient aeration through forced aeration or mechanical turning would guarantee the non-formation of anaerobic zones within the composting mixture, thereby reducing odor problems [9,86]. Importantly, and per statistical analysis, it absolutely was clear that aeration rate was the foremost important factor that could significantly affect the $NH_3$, $CH_4$, and $N_2O$ emissions [5,87]. Rosenfeld et al. [88] indicated that aeration reduced the concentrations of $NH_3$, $CH_2O_2$, and $CH_3COOH$ by 72%, 57%, and 11%, respectively, compared to the windrow. Additionally, Quiros et al. [89] reported a reduction in emissions' rates by five times when frequent turning was employed compared to non-turned treatments. However, it should be taken into account that an adequate aeration rate has to be applied that would maintain the biological activity and reduce the emissions' rates at the same time [5,90–92]. Applying higher aeration rates would reduce some emissions such as $CH_4$, but others such as $NH_3$ and $N_2O$ would be promoted. Additionally, higher aeration rates might render temperature evolution, thus decelerating the degradation of the organic matter as a consequence [5]. Results obtained by Chowdhury et al. [90] showed that low aeration rates were more practical in reducing GHGs' emissions. The same observation was reported by Zhang et al. [93] during composting of kitchen waste, where aeration rates of 0.1, 0.2, and 0.3 L (kg DM min)$^{-1}$ were studied and it was found that the lower aeration was more significant than the other two treatments. Additionally, it was indicated that intermittent aeration was better than continuous aeration in mitigating $CO_2$ emissions [9]. Nevertheless, Ermolaev et al. [71] indicated that aeration reduced the emissions of $CH_4$ and $NO_2$ regardless of the rate of aeration. Turning of the composted mixtures has a positive impact on reducing the rate of emissions also. This is because turning gives the chance for air exchange and releasing of different gases, as indicated in different studies [11,17]. The efficiency of aeration and turning in reducing emissions' rates was evaluated by Friedrich and Trois [10] during composting of garden waste. The study revealed that turning resulted in 8.14% higher GHGs than an aerated treatment. These findings prove that aeration is better than other treatments in reducing GHGs' emissions [94,95]. Another important element for maintaining the aerobic conditions is providing an optimal ratio of a bulking agent that provides proper structure and porosity for the composting mixture [31], meanwhile maintaining the heat and biological activity [96,97]. This will be deeply discussed in the following section.

### 3.6. pH Value

An optimum pH value between 6 and 8 (ideally 7) is recommended for microbial population during successful composting process. However, this parameter fluctuates, especially during the first stages of the process as organic acids are released due to organic matter degradation and, thus, the pH decreases. After that, a gradual increase in alkalinity occurs as a result of the phenolic and carboxyl groups' decomposition [4]. The increase in the pH level promotes ammonia release, by influencing the $NH_4^+$ to $NH_3$ equilibrium in spite of its directs influence on biological activity, as indicated [54,55]. However, these conditions are considered suitable for decreasing other emissions such as $H_2S$, which normally increases under low pH values [56]. These behaviors were documented by Gu et al. [98], where reducing the pH of compost resulted in reducing the cumulative $NH_3$ emissions and TN losses by 47.80% and 44.23%, while an increase in the emissions of volatile sulfur compounds and total sulfur losses was observed. More information and details about the effect of pH adjustment to mitigate the emission rates are provided in Section 4.4.

## 4. Mitigation Strategies

### 4.1. Providing Adequate Bulking Agent

The addition of some materials to organic wastes has proven its efficiency in improving air convection within the composting mixture, thereby reducing the amount of gases' emissions such as $CH_4$ and $N_2O$ from composting, since most of the degraded carbon would be released as $CO_2$ [11,28,90]. For instance, sawdust and straw for dairy manure composting resulted in an effective mitigation for $CH_4$ and $NH_3$ with ME values of 66.3% and 44.0%, but they may increase $CO_2$ emission [12,99]. Additionally, Li et al. [94] demonstrated that ammonia emission may well be mitigated by adding a mix of sucrose and straw powder at the start stage of a composting process [95]. Indeed, these materials facilitate the absorption and microbial assimilation of ammonium, which decreases $NH_3$ emissions [9,25,95].

### 4.2. Introducing Microorganism for Promoting Nitrification Process and Reducing $NH_3$ Emissions

This approach stands on the mineralization of organic nitrogen into ammonium nitrogen, which could be transformed into nitrate by nitrification and eventually to $N_2$ by denitrification, or the ammonium could even be also a fixed microbial protein under the action of fungi [25,90,100–103]. It was found that the introduction of mature compost rich in nitrifying microorganism to food wastes' composting was able to reduce $NH_3$ volatilization by 36% [104]. Nevertheless, and despite the capability of this approach in reducing $NH_3$ emission, regulating the denitrification process to reduce $N_2$ and $N_2O$ still represents a challenge for its successful application [5,103]. Additionally, the introduction of some exogenous microbial communities including CC-E (a complex bacterial community in which *Alcaligenes faecalis* is the main advantageous strain) and EM (Effective Microorganisms, a kind of commercial microbiological agent) for dairy manure composting reduced the potential for $NH_3$ emissions, with ME of 9.15% [104,105].

### 4.3. Vermicomposting

This composting approach demonstrated promising results in reducing the amounts of gaseous emissions including nitrous oxide, $CH_4$, $NH_3$, and others [95,106]. The decrease in emissions' rates is attributed to the reduction of anaerobic denitrification, due to the burrowing action of the earthworms [107]. Furthermore, the large specific surface area and loose texture in vermicomposting contribute to creating a strong adsorption capacity and, at last, reducing production of different emissions, among them the $NH_3$, where vermicomposting was able to mitigate $NH_3$ emission with a ME median value of 33.5% [15,25,108]. The loss of texture improves the aerobic conditions and,, therefore, the biodegradation of the organic matter as a consequence. In this regard, it was noticed that $CO_2$ emissions were increased, whereas a decrease in ammonia emissions and nitrous oxide was noticed as well as a sink of methane in treatments with earthworms [109,110]. Similar results were obtained by Chan et al. [108] and Velasco-Velasco et al. [111]. Combining pre-composting and vermicomposting with additions of reed straw and zeolite resulted also in a significant reduction of ammonia, nitrous oxide, and methane during composting of duck manure [95,111].

### 4.4. Using Different Additives

The addition of phosphogypsum results in decreasing the pH of the composting mixture. The high sulphide concentrations and acidic conditions due to the use of phosphogypsum could inhibit methanogenesis and the action of $N_2O$ reductase, thus reducing $CH_4$ and $N_2O$ emissions [12,25,112,113]. Additionally, adjustment of pH has been practiced to reduce the emissions of $NH_3$. About 55.7% of $NH_3$ emissions was decreased due to the reduction in volatilization when phosphogypsum was applied [114]. Additionally, the addition of both $K_2HPO_4/MgSO_4$ and $KH_2PO_4/MgSO_4$ as a pH buffer agent's additive contributed to reducing $NH_3$ emissions [100]. However, health risks due to high hydrogen sulphide concentrations have to be considered when this mitigation method is to be used [25,115]. Manure acidification significantly (up to 93%) decreased the emissions during storage and composting processes [116,117]. Excessive acidification (pH = 5), on the other hand, increased $N_2O$ emissions (18.6%) during composting. When manure was acidified to pH of 6, $N_2O$ (17.6%) and $CH_4$ (20%) emissions, as well as GHG emissions, represented as global warming potential (GWP) (9.6%) were reduced during composting [118]. The addition of calcium magnesium phosphate fertilizer (CaMgP) also demonstrated its effectiveness in reducing emissions' rates during the composting process [119]. In this regard, Zhang et al. [93] reported that CaMgP could reduce $H_2S$ emissions by 65%. Similar results were obtained when the effect of calcium magnesium phosphate fertilizer (CaMgP), biochar, and spent mushroom substrate (SMS) additives was investigated on compost maturity and gaseous emissions during pig manure composting. Ammonia ($NH_3$), hydrogen sulfide ($H_2S$), dimethyl sulfide ($Me_2S$), and dimethyl disulfide ($Me_2SS$) emissions could all be reduced using the three additives. However, when it came to reducing $NH_3$ emissions, the effect of adding CaMgP was the most noticeable (42.90%). CaMgP to $H_2S$ emission reduction was similar to SMS, which was 34.91% and 32.88%, respectively. The three additives had obvious emission reduction effects on $Me_2S$ and $Me_2SS$, all of which were greater than 50%. Adding SMS, on the other hand, reduced $N_2O$ emissions by 37.08% [120].

Struvite could also be used to reduce emissions as struvite crystallization enhances nitrogen (ammonium) conservation during composting, which thereby reduces $NH_3$ emissions [47,121]. However, this approach increases the salinity of the produced compost [5,94], but this limitation could be mitigated by using other additives like lime or zeolite [18,122]. In this regard, the addition of 10% of zeolite decreased the salinity to 2.8 mS $cm^{-1}$ and improved compost maturity; meanwhile, about 18% of $NH_3$ loss was achieved [122].

### 4.5. *Compressing and Covering*

This approach depends on reducing the amount of $O_2$ supplied to the mixture, thus lowering the microbial activity and ammonization, which reduce $CO_2$ and $NH_3$ emissions during the composting process [101,123]. Additionally, covering reduces gaseous diffusion into the air and enhances the absorption of some gas emissions. Analysis revealed that this approach could reach a mitigation efficiency of 10.1% for $CO_2$ and 24.3% for $NH_3$ emission. However, it should be noted that this approach would increase the anaerobic conditions that ultimately promote the production of $CH_4$ [15,25,107,109]. Different materials are used as a cover for composting mixture. These materials include sawdust, plastic, soil, paper waste, woodchip, wheat straw, peat, and zeolite, among others. Sawdust or straw has a good performance in absorption of $CO_2$ and $NH_3$, whereas plastic cover renders the gas exchange, which reduces the dissipation of the emissions [15,25,109,124,125]. Different forms of zeolite were used as a cover or even mixed with the composting mixture and proved higher efficiency in reducing emission compared to other cover materials with almost no effect on the microbial activity [5,93,104,126,127]. This material contributes to increasing the pH and initial $NH_3/NH_4^+$ concentration, which reduces $NH_3$ losses such that a reduction of 44–60% of the $NH_3$ was obtained during poultry manure composting [128]. Similar results were observed by Madrini et al. [126] in composting of leftover food. It should be noted that the type of zeolite and its percentage within the mixture affects the reduction rate of emissions [5,127].

### 4.6. *Biofiltration*

Biofilters, which depend on adsorption or biodegradation of pollutants, have proven their relative efficiency in reducing emissions from the composting process, especially with $NH_3$, where almost about 90% of this gas was reduced [25,129]. Actually, ammonia emissions in a composting process of organic fraction of municipal solid wastes varied between 18 to 150 g $NH_3 \cdot Mg^{-1}$ waste [130], while ammonia concentrations up to 700 mg $NH_3 \cdot m^{-3}$ have been reported in exhaust gases from sludge composting [4]. As documented by Pagans et al. [131], the biofilter achieved a global ammonia removal efficiency of 95.9% at a loading rate range of 846–67100 mg $NH_3 \cdot m^{-3}$ biofilter$\cdot h^{-1}$, whereas higher removal rates were seen when the waste gas had high $NH_3$ concentrations (more than 2000 mg $NH_3 \cdot m^{-3}$). However, this approach is more feasible compared to other technologies when it is used in closed systems with collection equipment [15]. Furthermore, the complexity and uncertainty measures in operating the system, as well as understanding the biodegradation process, are critical for optimal performance. [9]. Concerning $CH_4$, $CO_2$, and $N_2O$ emissions, the literature is lacking information about the efficiency of biofilter for treatment of these emissions [25].

### 4.7. Addition of Biochar

Biochar as an additive has been used in different research to mitigate the emissions resulting from composting processes [94,100,120,132,133]. This additive has been used as a sole material or mixed with other additives [134]. Noteworthy, under almost all studied conditions, promising results were obtained, despite the lack of clarity regarding its mechanism on promoting nitrogen assimilation and nitrification [5,90,102,135]. The change in nitrogen functional groups on the biochar surface was evidence for adsorption and microbial transformation of $NH_3/NH_4^+$ [136]. As indicated in several works, the biochar promoted microbial activity during the composting process, as it increases the nitrogen source and decreases toxicity of free $NH_3$ on the microbial activity [137]; hence, a high respiration rate as well as a fast decomposition of organic matter were recorded [135,137,138]. Additionally, this was associated with an increase in the temperature and $NO_3$ concentration along with a decrease in the pH and $NH_4^+$ concentrations [131,133]. Emissions of $NH_3$ and nitrogen losses were reduced by 64% and 52%, respectively, when biochar was mixed with poultry litters [101]. Similar results were observed when cornstalk biochar was used where cumulative $NH_3$ emissions were reduced by 24.8% [139]. The presence of the biochar boosted the activity of nitrifiers due to its high sorption capacity for gases and the high cation exchange capacity. According to Zhou et al. [140], adding modified biochar could significantly reduce $NH_3$ emissions by increasing the number of ammonia-oxidizing bacteria (AOB), inhibiting urease activity, and decreasing the abundance of nitrogen functional genes such as narG and nirS, facilitating the conversion of $NH^+{}_4$-N into $NO^-{}_3$-N and decreasing nitrogen loss. These conditions were responsible for promoting $N_2O$ reduction up to 59.8% [141]. The effects of bamboo charcoal (BC) and bamboo vinegar (BV) on lowering $NH_3$ and $N_2O$ emissions during aerobic composting (Wheat straw and pig manure) revealed that both BC and BV enhanced nitrogen conversion and compost quality, with the combination BC + BV treatment achieving the greatest results. The BC, BV, and BC + BV treatments decreased $NH_3$ emissions by 14.35%, 17.90%, and 29.83%, respectively, and the $N_2O$ emissions by 44.83%, 55.96%, and 74.53%. BC and BV reduced the $NH_3$ and $N_2O$ emissions during composting [142]. Similarly, Biochar (BC) and bean dregs' (BD) effects on nitrifiers and denitrifiers, as well as contributions to $NH_3$ and $N_2O$ emissions, were investigated by Yang et al. [143]. When comparing the BD + BC treatment to the BD treatment, the highest value of $NH_3$ and $N_2O$ emission was reduced by 32.92% and 46.61%, respectively. The number and structure of nitrogen functional genes were shown to be closely related to the synthesis of $NH_3$ and $N_2O$ in the study. In this case, it was discovered that BD + BC enhanced the abundance of the AOB amoA gene, resulting in a reduction in $NH_3$ emission. The presence of nirS was more closely linked to the presence of $N_2O$. When compared to the BD treatment, the abundance of nirS in the BD + BC treatment was reduced by 18.93%, lowering $N_2O$ emissions after composting. Furthermore, the nosZ-type gene was the most functional denitrification bacterial community to influence $N_2O$ emissions. [143]. Noteworthy, when biochar is to be used, it is important to keep in mind that its characteristics have a major role on its efficiency.

## 5. Conclusions

Composting is a favorable technology to treat organic waste, but gaseous emissions are an issue of major concern for its development. Among them, GHG emissions are an important problem as they are responsible for the global warming effect. Carbon dioxide is not often considered, as it is considered biogenic. However, methane and nitrous oxide, related to anaerobic and anoxic conditions, must be accounted for when analyzing any composting process. Another important point is the release in the form of gaseous emissions of a vast family of compounds such as VOCs. These gases can be harmful, possess negative impacts, and, especially, are responsible for unpleasant odors. The origin of these gases is double (they can come from the substrate or be biologically or even chemically formed during the process) and they need the development of mitigation strategies based on relatively consolidated technologies (such as biofiltration) or new approaches, such as the use of materials as biochar. However, there is still a lack of reliable and full-scale data from composting emissions to have consistent mitigation strategies.

**Author Contributions:** Writing—original draft preparation, T.S.; writing—review and editing, A.S. All authors have read and agreed to the published version of the manuscript.

**Funding:** This research received no external funding.

**Acknowledgments:** Sayara would like to thank the Palestine Technical University for the administrative support.

**Conflicts of Interest:** The authors declare no conflict of interest.

## References

1. Iqbal, A.; Liu, X.; Chen, G. Municipal solid waste: Review of best practices in application of life cycle assessment and sus-tainable management techniques. *Sci. Total Environ.* **2020**, *729*, 138622. [CrossRef]
2. Kaza, S.; Yao, L.; Bhada-Tata, P.; Van Woerden, F. *What a Waste 2.0: A Global Snapshot of Solid Waste Management to 2050*; Urban Development Series; World Bank Publications: Washington, DC, USA, 2018. [CrossRef]
3. Bogner, J.; Pipatti, R.; Hashimoto, R.; Diaz, C.; Mareckova, K.; Diaz, L.; Kjeldsen, P.S.; Faaij, A.; Gao, Q.; Zhang, T.; et al. Mitigation of global GHG emissions from waste: Conclusions and strategies from the Intergovernmental Panel on Climate Change (IPCC) Fourth Assessment Report. Working Group III (Mitigation). *Waste Manag. Res.* **2008**, *26*, 11–13. [CrossRef] [PubMed]
4. Haug, R. *The Practical Handbook of Compost Engineering*; Lewis Publishers: Boca Raton, FL, USA, 1993. [CrossRef]
5. Wang, S.; Zeng, Y. Ammonia emission mitigation in food waste composting: A review. *Bioresour. Technol.* **2018**, *248*, 13–19. [CrossRef] [PubMed]
6. Hodge, K.L.; Levis, J.W.; Decarolis, J.F.; Barlaz, M.A. Systematic Evaluation of Industrial, Commercial, and Institutional Food Waste Management Strategies in the United States. *Environ. Sci. Technol.* **2016**, *50*, 8444–8452. [CrossRef] [PubMed]
7. Sayara, T.; Basheer-Salimia, R.; Hawamde, F.; Sanchez, A. Recycling of Organic Wastes through Composting: Process Per-formance and Compost Application in Agriculture. *Agronomy* **2020**, *10*, 1838. [CrossRef]
8. Varma, V.S.; Kalamdhad, A.S. Stability and microbial community analysis during rotary drum composting of vegetable waste. *Int. J. Recycl. Org. Waste Agric.* **2014**, *3*, 52. [CrossRef]
9. Dhamodharan, K.; Varma, V.S.; Veluchamy, C.; Pugazhendhi, A.; Rajendran, K. Emission of volatile organic compounds from composting: A review on assessment, treatment and perspectives. *Sci. Total. Environ.* **2019**, *695*, 133725. [CrossRef]
10. Friedrich, E.; Trois, C. GHG emission factors developed for the recycling and composting of municipal waste in South African municipalities. *Waste Manag.* **2013**, *33*, 2520–2531. [CrossRef]
11. Luo, W.; Yuan, J.; Luo, Y.M.; Li, G.X.; Nghiem, L.; Price, W.E. Effects of mixing and covering with mature compost on gaseous emissions during composting. *Chemosphere* **2014**, *117*, 14–19. [CrossRef]
12. Hao, X.; Chang, C.; Larney, F.J. Carbon, nitrogen balances and greenhouse gas emission during cattle feedlot manure composting. *J. Environ. Qual.* **2004**, *33*, 37–44. [CrossRef]
13. Colón, J.; Cadena, E.; Pognani, M.; Barrena, R.; Sánchez, A.; Font, X.; Artola, A. Determination of the energy and environ-mental burdens associated with the biological treatment of source-separated municipal solid wastes. *Energy Environ. Sci.* **2012**, *5*, 5731–5741. [CrossRef]
14. Komilis, D.P.; Ham, R.K.; Park, J.K. Emission of volatile organic compounds during composting of municipal solid wastes. *Water Res.* **2004**, *38*, 1707–1714. [CrossRef] [PubMed]
15. Wang, Y.; Li, X.; Yang, J.; Tian, Z.; Sun, Q.; Xue, W.; Dong, H. Mitigating Greenhouse Gas and Ammonia Emissions from Beef Cattle Feedlot Production: A System Meta-Analysis. *Environ. Sci. Technol.* **2018**, *52*, 11232–11242. [CrossRef]
16. Zhu, L.; Yang, H.; Zhao, Y.; Kang, K.; Liu, Y.; He, P.; Wu, Z.; Wei, Z. Biochar combined with montmorillonite amendments increase bioavailable organic nitrogen and reduce nitrogen loss during composting. *Bioresour. Technol.* **2019**, *294*, 122224. [CrossRef] [PubMed]
17. Jiang, T.; Li, G.; Tang, Q.; Ma, X.; Wang, G.; Schuchardt, F. Effects of aeration method and aeration rate on greenhouse gas emissions during composting of pig feces in pilot scale. *J. Environ. Sci.* **2015**, *31*, 124–132. [CrossRef]
18. Wang, X.; Selvam, A.; Wong, J.W. Influence of lime on struvite formation and nitrogen conservation during food waste composting. *Bioresour. Technol.* **2016**, *217*, 227–232. [CrossRef] [PubMed]

19. Yang, F.; Li, G.; Shi, H.; Wang, Y. Effects of phosphogypsum and superphosphate on compost maturity and gaseous emis-sions during kitchen waste composting. *Waste Manag.* **2015**, *36*, 70–76. [CrossRef]
20. Friedrich, E.; Trois, C. Quantification of greenhouse gas emissions from waste management processes for municipalities—A comparative review focusing on Africa. *Waste Manag.* **2011**, *31*, 1585–1596. [CrossRef] [PubMed]
21. Saer, A.; Lansing, S.; Davitt, N.H.; Graves, R.E. Life cycle assessment of a food waste composting system: Environmental impact hotspots. *J. Clean. Prod.* **2013**, *52*, 234–244. [CrossRef]
22. Wang, K.; Nakakubo, T. Comparative assessment of waste disposal systems and technologies with regard to greenhouse gas emissions: A case study of municipal solid waste treatment options in China. *J. Clean. Prod.* **2020**, *260*, 120827. [CrossRef]
23. Delgado-Rodríguez, M.; Ruiz-Montoya, M.; Giráldez, I.; Cabeza, I.O.; Lopez, R.; Diaz, M.J. Effect of control parameters on emitted volatile compounds in municipal solid waste and pine trimmings composting. *J. Environ. Sci. Heal. Part. A* **2010**, *45*, 855–862. [CrossRef]
24. Font, X.; Artola, A.; Sánchez, A. Detection, Composition and Treatment of Volatile Organic Compounds from Waste Treatment Plants. *Sensors* **2011**, *11*, 4043–4059. [CrossRef]
25. Ba, S.; Qu, Q.; Zhang, K.; Groot, J.C.J. Meta-analysis of greenhouse gas and ammonia emissions from dairy manure com-posting. *Biosyst. Eng.* **2020**, *193*, 126–137. [CrossRef]
26. Shah, G.M.; Groot, J.C.J.; Oenema, O.; Lantinga, E.A. Covered storage reduces losses and improves crop utilisation of nitrogen from solid cattle manure. *Nutr. Cycl. Agroecosyst.* **2012**, *94*, 299–312. [CrossRef]
27. IPCC. Climate change 2013: The physical science Basis. In *Contribution of Working Group I to the Fifth Assessment Report of the Intergovernmental Panel on Climate Change*; Intergovernmental Panel on Climate Change (IPCC): Geneva, Switzerland, 2013.
28. Maeda, K.; Hanajima, D.; Morioka, R.; Toyoda, S.; Yoshida, N.; Osada, T. Mitigation of greenhouse gas emission from the cattle manure composting process by use of a bulking agent. *Soil Sci. Plant. Nutr.* **2013**, *59*, 96–106. [CrossRef]
29. Mulbry, W.; Ahn, H. Greenhouse gas emissions during composting of dairy manure: Influence of the timing of pile mixing on total emissions. *Biosyst. Eng.* **2014**, *126*, 117–122. [CrossRef]
30. Sommer, S.G. Effect of composting on nutrient loss and nitrogen availability of cattle deep litter. *Eur. J. Agron.* **2001**, *14*, 123–133. [CrossRef]
31. Sánchez, A.; Artola, A.; Font, X.; Gea, T.; Barrena, R.; Gabriel, D.; Sanchez-Monedero, M.; Roig, A.; Cayuela, M.L.; Mondini, C. Greenhouse gas emissions from organic waste composting. *Environ. Chem. Lett.* **2015**, *13*, 223–238. [CrossRef]
32. Turan, N.G.; Akdemir, A.; Ergun, O.N. Emission of Volatile Organic Compounds during Composting of Poultry Litter. *Water, Air, Soil Pollut.* **2007**, *184*, 177–182. [CrossRef]
33. Sánchez-Monedero, M.A.; Sánchez-García, M.; Alburquerque, J.A.; Cayuela, M.L. Biochar reduces volatile organic com-pounds generated during chicken manure composting. *Bioresour Technol.* **2019**, *288*, 121584. [CrossRef] [PubMed]
34. Maulini-Duran, C.; Artola, A.; Font, X.; Sánchez, A. A systematic study of the gaseous emissions from biosolids composting: Raw sludge versus anaerobically digested sludge. *Bioresour. Technol.* **2013**, *147*, 43–51. [CrossRef] [PubMed]
35. Akdeniz, N.; Koziel, J.A.; Ahn, H.K.; Glanville, T.D.; Crawford, B.P.; Raman, D.R. Laboratory scale evaluation of volatile organic compound emissions as indication of swine carcass degradation inside biosecure composting units. *Bioresour. Technol.* **2010**, *101*, 71–78. [CrossRef] [PubMed]
36. Sánchez-Monedero, M.A.; Fernández-Hernández, A.; Higashikawa, F.S.; Cayuela, M.L. Relationships between emitted volatile organic compounds and their concentration in the pile during municipal solid waste composting. *Waste Manag.* **2018**, *79*, 179–187. [CrossRef]
37. Wang, Y.-J.; Xing, Z.-X.; Zhang, X.-F.; Hou, Z.-G.; Zhao, X.-S.; Dou, S.; Zhou, M.-P. On-site Detection of Volatile Organic Compounds During Composting Treatment of Livestock and Poultry Manure by GC-MS. *Chin. J. Anal. Chem.* **2013**, *40*, 899–903. [CrossRef]
38. Rincón, C.A.; De Guardia, A.; Couvert, A.; Le Roux, S.; Soutrel, I.; Daumoin, M.; Benoist, J.C. Chemical and odor charac-terization of gas emissions released during composting of solid wastes and digestates. *J. Environ. Manag.* **2019**, *233*, 39–53. [CrossRef] [PubMed]
39. Kumar, A.; Alaimo, C.P.; Horowitz, R.; Mitloehner, F.M.; Kleeman, M.J.; Green, P.G. Volatile organic compound emissions from green waste composting: Characterization and ozone formation. *Atmos. Environ.* **2011**, *45*, 1841–1848. [CrossRef]
40. González, D.; Colón, J.; Sánchez, A.; Gabriel, D. A systematic study on the VOCs characterization and odour emissions in a full-scale sewage sludge composting plant. *J. Hazard. Mater.* **2019**, *373*, 733–740. [CrossRef]
41. Goldstein, N. Getting to know the odor compounds. *Biocycle* **2002**, *43*, 42–44.
42. Nag, M.; Shimaoka, T.; Komiya, T. Nitrous oxide production during nitrification from organic solid waste under tempera-ture and oxygen conditions. *Environ. Technol.* **2016**, *37*, 2890–2897. [CrossRef]
43. Hellebrand, H.J. Emission of Nitrous Oxide and other Trace Gases during Composting of Grass and Green Waste. *J. Agric. Eng. Res.* **1998**, *69*, 365–375. [CrossRef]
44. Amlinger, F.; Peyr, S.; Cuhls, C. Greenhouse gas emissions from composting and mechanical biological treatment. *Waste Manag. Res.* **2008**, *26*, 47–60. [CrossRef] [PubMed]
45. González, D.; Guerra, N.; Colón, J.; Gabriel, D.; Ponsa, S.; Sánchez, A. Characterization of the Gaseous and Odour Emissions from the Composting of Conventional Sewage Sludge. *Atmosphere* **2020**, *11*, 211. [CrossRef]

46. Pan, J.; Cai, H.; Zhang, Z.; Liu, H.; Li, R.; Mao, H.; Awasthi, M.K.; Wang, Q.; Zhai, L. Comparative evaluation of the use of acidic additives on sewage sludge composting quality improvement, nitrogen conservation, and greenhouse gas reduction. *Bioresour. Technol.* **2018**, *270*, 467–475. [CrossRef]
47. Yuan, J.; Chadwick, D.; Zhang, D.; Li, G.; Chen, S.; Luo, W.; Du, L.; He, S.; Peng, S. Effects of aeration rate on maturity and gaseous emissions during sewage sludge composting. *Waste Manag.* **2016**, *56*, 403–410. [CrossRef]
48. Lou, X.F.; Nair, J. The impact of landfilling and composting on greenhouse gas emissions—A review. *Bioresour. Technol.* **2009**, *100*, 3792–3798. [CrossRef]
49. Banar, M.; Cokaygil, Z.; Ozkan, A. Life cycle assessment of solid waste management options for Eskisehir, Turkey. *Waste Manag.* **2009**, *29*, 54–62. [CrossRef] [PubMed]
50. Güereca, L.P.; Gassó, S.; Baldasano, J.M.; Jiménez-Guerrero, P. Life cycle assessment of two biowaste management systems for Barcelona, Spain. *Resour. Conserv. Recycl.* **2006**, *49*, 32–48. [CrossRef]
51. Yay, A.S.E. Application of life cycle assessment (LCA) for municipal solid waste management: A case study of Sakarya. *J. Clean. Prod.* **2015**, *94*, 284–293. [CrossRef]
52. Komakech, A.J.; Sundberg, C.; Jönsson, H.; Vinnerås, B. Life cycle assessment of biodegradable waste treatment systems for sub-Saharan African cities. *Resour. Conserv. Recycl.* **2015**, *99*, 100–110. [CrossRef]
53. Bohacz, J. Changes in mineral forms of nitrogen and sulfur and enzymatic activities during composting of lignocellulosic waste and chicken feathers. *Environ. Sci. Pollut. Res.* **2019**, *26*, 10333–10342. [CrossRef]
54. Zhu, P.; Shen, Y.; Pan, X.; Dong, B.; Zhou, J.; Zhang, W.; Li, X. Reducing odor emissions from feces aerobic composting: Additives. *RSC Adv.* **2021**, *11*, 15977–15988. [CrossRef]
55. Liang, Y.; Leonard, J.J.; Feddes, J.J.; McGill, W.B. A SIMULATION MODEL OF AMMONIA VOLATILIZATION IN COMPOSTING. *Trans. ASAE* **2004**, *47*, 1667–1680. [CrossRef]
56. Wang, Y.; Liu, S.J.; Xue, W.T.; Guo, H.; Li, X.R.; Zuo, G.Y.; Zhao, T.K.; Dong, H.M. The Characteristics of Carbon, Ni-trogen and Sulfur Transformation During Cattle Manure Composting-Based on Different Aeration Strategies. *Int. J. Environ. Res. Public Health.* **2019**, *16*, 3930. [CrossRef] [PubMed]
57. Chen, J.; Chen, T.-B.; Gao, D.; Lei, M.; Zheng, G.-D.; Liu, H.-T.; Guo, S.-L.; Cai, L. Reducing $H_2S$ production by $O_2$ feedback control during large-scale sewage sludge composting. *Waste Manag.* **2011**, *31*, 65–70. [CrossRef] [PubMed]
58. Amon, B.; Amon, T.; Boxberger, J.; Alt, C. Emissions of $NH_3$, $N_2O$ and $CH_4$ from dairy cows housed in a farmyard manure tying stall (housing, manure storage, manure spreading). *Nutr. Cycl. Agroecosyst.* **2001**, *60*, 103–113. [CrossRef]
59. Fillingham, M.A.; VanderZaag, A.C.; Burtt, S.; Balde, H.; Ngwabie, N.M.; Smith, W.; Hakami, A.; Wagner-Riddle, C.; Bittman, S.; MacDonald, D. Greenhouse gas and ammonia emissions from production of compost bedding on a dairy farm. *Waste Manag.* **2017**, *70*, 45–52. [CrossRef] [PubMed]
60. Guest, G.; Smith, W.; Grant, B.; VanderZaag, A.; Desjardins, R.; McConkey, B. A comparative life cycle assessment high-lighting the trade-offs of a liquid manure separatorcomposter in a Canadian dairy farm system. *J. Clean. Prod.* **2017**, *143*, 824–835. [CrossRef]
61. Ahn, H.K.; Mulbry, W.; White, J.W.; Kondrad, S.L. Pile mixing increases greenhouse gas emissions during composting of dairy manure. *Bioresour. Technol.* **2011**, *102*, 2904–2909. [CrossRef] [PubMed]
62. Arriaga, H.; Viguria, M.; López, D.M.; Merino, P. Ammonia and greenhouse gases losses from mechanically turned cattle manure windrows: A regional composting network. *J. Environ. Manag.* **2017**, *203*, 557–563. [CrossRef]
63. El Kader, N.A.; Robin, P.; Paillat, J.-M.; Leterme, P. Turning, compacting and the addition of water as factors affecting gaseous emissions in farm manure composting. *Bioresour. Technol.* **2007**, *98*, 2619–2628. [CrossRef]
64. Hao, X.; Chang, C.; Larney, F.J.; Travis, G.R. Greenhouse Gas Emissions during Cattle Feedlot Manure Composting. *J. Environ. Qual.* **2001**, *30*, 376–386. [CrossRef] [PubMed]
65. He, Y.; Inamori, Y.; Mizuochi, M.; Kong, H.; Iwami, N.; Sun, T. Nitrous Oxide Emissions from Aerated Composting of Organic Waste. *Environ. Sci. Technol.* **2001**, *35*, 2347–2351. [CrossRef]
66. Burgin, A.J.; Groffman, P.M. Soil $O_2$ controls denitrification rates and $N_2O$ yield in a riparian wetland. *J. Geophys. Res Biogeosci.* **2012**, *117*. [CrossRef]
67. Morley, N.; Baggs, E.M.; Dörsch, P.; Bakken, L. Production of NO, $N_2O$ and $N_2$ by extracted soil bacteria, regulation by $NO_2$- and $O_2$ concentrations. *FEMS Microbiol. Ecol.* **2008**, *65*, 102–112. [CrossRef]
68. Saggar, S.; Jha, N.; Deslippe, J.; Bolan, N.S.; Luo, J.; Giltrap, D.L.; Kim, D.-G.; Zaman, M.; Tillman, R. Denitrification and $N_2O$:$N_2$ production in temperate grasslands: Processes, measurements, modelling and mitigating negative impacts. *Sci. Total. Environ.* **2013**, *465*, 173–195. [CrossRef] [PubMed]
69. Parkinson, R.; Gibbs, P.; Burchett, S.; Misselbrook, T. Effect of turning regime and seasonal weather conditions on nitrogen and phosphorus losses during aerobic composting of cattle manure. *Bioresour. Technol.* **2004**, *91*, 171–178. [CrossRef]
70. Szanto, G.L.; Hamelers, H.V.M.; Rulkens, W.H.; Veeken, A.H.M. $NH_3$, $N_2O$ and $CH_4$ emissions during passively aerated composting of straw-rich pig manure. *Bioresour. Technol.* **2007**, *98*, 2659–2670. [CrossRef]
71. Ermolaev, E.; Pell, M.; Smårs, S.; Sundberg, C.; Jönsson, H. Greenhouse gas emission from covered windrow composting with controlled ventilation. *Waste Manag. Res.* **2012**, *30*, 155–160. [CrossRef]
72. Pattey, E.; Trzcinski, M.K.; Desjardins, R.L. Quantifying the reduction of greenhouse gas emissions as a result of com-posting dairy and beef cattle manure. *Nutr. Cycling Agroecosyst.* **2005**, *72*, 173–187. [CrossRef]

73. Sánchez-Monedero, M.A.; Serramiá, N.; Civantos, C.G.; Fernández-Hernández, A.; Roig, A. Greenhouse gas emissions during composting of two-phase olive mill wastes with different agroindustrial by-products. *Chemosphere* **2010**, *81*, 18–25. [CrossRef]
74. Yamulki, S. Effect of straw addition on nitrous oxide and methane emissions from stored farmyard manures. *Agric. Ecosyst. Environ.* **2006**, *112*, 140–145. [CrossRef]
75. Yang, F.; Li, G.; Zang, B.; Zhang, Z. The maturity and $CH_4$, $N_2O$, $NH_3$ emissions from vermicomposting with agricultural waste. *Compost Sci. Util.* **2017**, *25*, 262–271. [CrossRef]
76. Qin, L.; Shen, Y.; Li, G.; Hu, J. C matter change of composting with different C/N. *J. Agro-Environ. Sci.* **2010**, *29*, 1388–1393.
77. González, D.G.; Colón, J.; Gabriel, D.; Sánchez, A. The effect of the composting time on the gaseous emissions and the compost stability in a full-scale sewage sludge composting plant. *Sci. Total. Environ.* **2019**, *654*, 311–323. [CrossRef]
78. Xie, J.; Li, Y.; Dong, H.; Huang, H.; Yu, Q. 2003 Influence of different factors on greenhouse gas emissions in composting of layer-hen manure with closed composting bins. *Trans. Chin. Soc. Agric. Eng.* **2003**, *19*, 192–195.
79. Hwang, S.; Hanaki, K. Effects of oxygen concentration and moisture content of refuse on nitrification, denitrification and nitrous oxide production. *Bioresour. Technol.* **2000**, *71*, 159–165. [CrossRef]
80. Yan, Y.; Wang, D.; Zheng, J. Advances in effects of biochar on the soil $N_2O$ and $CH_4$ emissions. *Chin. Agric. Sci. Bull.* **2013**, *29*, 140–146.
81. Hao, S. Process of Windrow and Trough Type Composting in Cold Region. Master's Thesis, Heilongjiang Bayi Agricultural University, Daqing, China, 2011.
82. Tsutsui, H.; Fujiwara, T.; Matsukawa, K.; Funamizu, N. Nitrous oxide emission mechanisms during intermittently aerated composting of cattle manure. *Bioresour. Technol.* **2013**, *141*, 205–211. [CrossRef]
83. Huang, G.F.; Wong, J.W.C.; Wu, Q.T.; Nagar, B.B. Effect of C/N on composting of pig manure with sawdust. *Waste Manag.* **2004**, *24*, 805–813. [CrossRef]
84. Pagans, E.; Font, X.; Sanchez, A. Emission of volatile organic compounds from composting of different solid wastes: Abatement by biofiltration. *J. Hazard. Mater.* **2006**, *131*, 179–186. [CrossRef]
85. Tong, B.; Wang, X.; Wang, S.; Ma, L.; Ma, W. Transformation of nitrogen and carbon during composting of manure litter with different methods. *Bioresour. Technol.* **2019**, *293*, 122046. [CrossRef] [PubMed]
86. He, P.; Wei, S.; Shao, L.; Lü, F. Emission potential of volatile sulfur compounds (VSCs) and ammonia from sludge compost with different bio-stability under various oxygen levels. *Waste Manag.* **2017**, *73*, 113–122. [CrossRef] [PubMed]
87. Jiang, T.; Schuchardt, F.; Li, G.; Guo, R.; Zhao, Y. Effect of C/N ratio, aeration rate and moisture content on ammonia and greenhouse gas emission during the composting. *J. Environ. Sci.* **2011**, *23*, 1754–1760. [CrossRef]
88. Rosenfeld, P.; Grey, M.; Sellew, P. 2004 Measurement of biosolids compost odor emissions from a windrow, static pile, and biofilter. *Water Environ. Res.* **2004**, *76*, 310–315. [CrossRef]
89. Quiros, D.C.; Yoon, S.; Dwyer, H.A.; Collins, J.F.; Zhu, Y.; Huai, T. Measuring particulate matter emissions during parked active diesel particulate filter regeneration of heavy-duty diesel trucks. *J. Aerosol Sci.* **2014**, *73*, 48–62. [CrossRef]
90. Chowdhury, M.A.; de Neergaard, A.; Jensen, L.S. Potential of aeration flow rate and bio-char addition to reduce greenhouse gas and ammonia emissions during manure composting. *Chemosphere* **2014**, *97*, 16–25. [CrossRef] [PubMed]
91. Gao, M.C.; Li, B.; Yu, A.; Liang, F.; Yang, L.; Sun, Y. The effect of aeration rate on forced-aeration composting of chicken manure and sawdust. *Bioresour. Technol.* **2010**, *101*, 1899–1903. [CrossRef] [PubMed]
92. Talib, A.T.; Mokhtar, M.N.; Baharuddin, A.S.; Sulaiman, A. Effects of aeration rate on degradation process of oil palm empty fruit bunch with kinetic-dynamic modeling. *Bioresour. Technol.* **2014**, *169*, 428–438. [CrossRef]
93. Zhang, J.; Sui, Q.; Li, K.; Chen, M.; Tong, J.; Qi, L.; Wei, Y. Influence of natural zeolite and nitrification inhibitor on organics degradation and nitrogen transformation during sludge composting. *Environ. Sci. Pollut. Res.* **2017**, *24*, 9122. [CrossRef]
94. Wang, X.; Selvam, A.; Chan, M.; Wong, J.W.C. Nitrogen conservation and acidity control during food wastes com-posting through struvite formation. *Bioresour. Technol.* **2013**, *147*, 17–22. [CrossRef]
95. Lim, S.L.; Lee, L.H.; Wu, T.Y. Sustainability of using composting and vermicomposting technologies for organic solid waste biotransformation: Recent overview, greenhouse gases emissions and economic analysis. *J. Clean. Prod.* **2016**, *111*, 262–278. [CrossRef]
96. Jolanun, B.; Towprayoon, S. Novel bulking agent from clay residue for food waste composting. *Bioresour. Technol.* **2010**, *101*, 4484–4490. [CrossRef]
97. Iqbal, M.K.; Shafiq, T.; Ahmed, K. Characterization of bulking agents and its effects on physical properties of compost. *Bioresour. Technol.* **2010**, *101*, 1913–1919. [CrossRef] [PubMed]
98. Gu, W.J.; Sun, W.; Lu, Y.S.; Li, X.; Xu, P.Z.; Xie, K.Z.; Sun, L.L.; Wu, H.T. Effect of Thiobacillus thioparus 1904 and sulphur addition on odour emission during aerobic composting. *Bioresour. Technol.* **2018**, *249*, 254–260. [CrossRef]
99. Barrington, S.; Choinière, D.; Trigui, M.; Knight, W. Compost convective airflow under passive aeration. *Bioresour. Technol.* **2003**, *86*, 259–266. [CrossRef]
100. Li, S.; Huang, G.-H.; An, C.-J.; Yu, H. Effect of different buffer agents on in-vessel composting of food waste: Performance analysis and comparative study. *J. Environ. Sci. Heal. Part. A* **2013**, *48*, 772–780. [CrossRef] [PubMed]
101. Chadwick, D.R. Emissions of ammonia, nitrous oxide and methane from cattle manure heaps: Effect of compaction and covering. *Atmos. Environ.* **2005**, *39*, 787–799. [CrossRef]

102. Chowdhury, M.A.; de Neergaard, A.; Jensen, L.S. Composting of solids separated from anaerobically digested animal manure: Effect of different bulking agents and mixing ratios on emissions of greenhouse gases and ammonia. *Biosys. Eng.* **2014**, *124*, 63–77. [CrossRef]
103. Steiner, C.; Das, K.C.; Melear, N.; Lakly, D. Reducing Nitrogen Loss during Poultry Litter Composting Using Biochar. *J. Environ. Qual.* **2010**, *39*, 1236–1242. [CrossRef]
104. Al-Jabi, L.F.; Halalsheh, M.M.; Badarneh, D.M. Conservation of ammonia during food waste composting. *Environ. Technol.* **2008**, *29*, 1067–1073. [CrossRef]
105. Zeng, Y.; De Guardia, A.; Dabert, P. Imporve composting as a post-treatment of anaerobic digestate. *Bioresour. Technol.* **2016**, *201*, 293–303. [CrossRef]
106. Chen, W.; Yan, L.; Gao, Y.; Bao, J.; Wang, Y.; Sun, Z.; Wang, W. The removal characteristics and diversity of a microbial community capable of ammonia removal from compost. *Ann. Microbiol.* **2016**, *66*, 635–642. [CrossRef]
107. Shan, J.; Shao, X. Nitrogen preserving and deodorizing technology in high temperature composting of cow manure. *Environ. Sci. Technol.* **2008**, *31*, 47–50.
108. Chan, Y.C.; Sinha, R.K.; Wang, W.J. Emission of greenhouse gases from home aerobic composting, anaerobic digestion and vermicomposting of household wastes in Brisbane (Australia). *Waste Manag. Res.* **2010**, *29*, 540–548. [CrossRef]
109. Chen, R.; Wang, Y.; Wang, W.; Wei, S.; Jing, Z.; Lin, X. $N_2O$ emissions and nitrogen transformation during windrow com-posting of dairy manure. *J. Environ. Manag.* **2015**, *160*, 121–127. [CrossRef]
110. Robin, P.; Germain, P.; Lecomte, M.; Landrain, B.; Li, Y.; Cluzeau, D. Earthworm effects on gaseous emissions during vermifiltration of pig fresh slurry. *Bioresour. Technol.* **2011**, *102*, 3679–3686.
111. Velasco-Velasco, J.; Parkinson, R.; Kuri, V. Ammonia emissions during vermicomposting of sheep manure. *Bioresour. Technol.* **2011**, *102*, 10959–10964. [CrossRef] [PubMed]
112. Wang, J.; Hu, Z.; Xu, X.; Jiang, X.; Zheng, B.; Liu, X.; Pan, X.; Kardol, P. Emissions of ammonia and greenhouse gases during combined pre-composting and vermicomposting of duck manure. *Waste Manag.* **2014**, *34*, 1546–1552. [CrossRef]
113. Liu, B.; Mørkved, P.T.; Frostegård, A.; Bakken, L.R. Denitrification gene pools, transcription and kinetics of NO, $N_2O$ and $N_2$ production as affected by soil pH. *FEMS Microbiol. Ecol.* **2010**, *72*, 407–417. [CrossRef] [PubMed]
114. Tubail, K.; Chen, L.; Michel, F.C., Jr.; Keener, H.M.; Rigot, J.F.; Klingman, M.; Kost, D.; Dick, W.A. Gypsum additions reduce ammonia nitrogen losses during composting of dairy manure and biosolids. *Compost Sci. Util.* **2008**, *16*, 285–293. [CrossRef]
115. Wang, Q.; Wang, Z.; Awasthi, M.K.; Jiang, Y.; Li, R.; Ren, X.; Zhao, J.; Shen, F.; Wang, M.; Zhang, Z. Evaluation of medical stone amendment for the reduction of nitrogen loss and bioavailability of heavy metals during pig manure composting. *Bioresour. Technol.* **2016**, *220*, 297–304. [CrossRef] [PubMed]
116. Hou, Y.; Velthof, G.L.; Oenema, O. Mitigation of ammonia, nitrous oxide and methane emissions from manure management chains: A meta-analysis and integrated assessment. *Glob. Chang. Biol.* **2015**, *21*, 1293–1312. [CrossRef]
117. Ti, C.; Xia, L.; Chang, S.X.; Yan, X. Potential for mitigating global agricultural ammonia emission: A meta-analysis. *Environ. Pollut.* **2019**, *245*, 141–148. [CrossRef]
118. Cao, Y.; Wang, X.; Liu, L.; Velthof, G.L.; Misselbrook, T.; Bai, Z.; Ma, L. Acidification of manure reduces gaseous emissions and nutrient losses from subsequent composting process. *J. Environ. Manag.* **2020**, *264*, 110454. [CrossRef] [PubMed]
119. Li, Y.; Luo, W.; Li, G.; Wang, K.; Gong, X. Performance of phosphogypsum and calcium magnesium phosphate fertilizer for nitrogen conservation in pig manure composting. *Bioresour. Technol.* **2018**, *250*, 53–59. [CrossRef]
120. Liu, Y.; Ma, R.; Li, D.; Qi, C.; Han, L.; Chen, M.; Fu, F.; Yuan, J.; Li, G. Effects of calcium magnesium phosphate fertilizer, biochar and spent mushroom substrate on compost maturity and gaseous emissions during pig manure composting. *J. Environ. Manag.* **2020**, *267*, 110649. [CrossRef] [PubMed]
121. Jiang, T.; Ma, X.; Tang, Q.; Yang, J.; Li, G.; Schuchardt, F. Combined use of nitrification inhibitor and struvite crystallization to reduce the $NH_3$ and $N_2O$ emissions during composting. *Bioresour. Technol.* **2016**, *217*, 210–218. [CrossRef] [PubMed]
122. Chan, M.T.; Selvam, A.; Wong, J.W. Reducing nitrogen loss and salinity during 'struvite' food waste composting by zeolite amendment. *Bioresour. Technol.* **2016**, *200*, 838–844. [CrossRef]
123. Jungbluth, T.; Hartung, E.; Brose, G. Greenhouse gas emissions from animal houses and manure stores. *Nutr. Cycl. Agroecosystems* **2001**, *60*, 133–145. [CrossRef]
124. Huang, G.; Fang, C.; Ma, S.; Han, L. Storage stability of micro-aerobic coupling functional membrane and gas emission reduction of dairy manure. *Trans. Chin. Soc. Agric. Mach.* **2018**, *49*, 335–341.
125. Zhu, X.; Dong, W.; Wang, H.; Yan, C.; Liu, H.; Liu, E. Effects of cattle manure composting methods on greenhouse gas and ammonia emissions. *Trans. Chin. Soc. Agric. Eng.* **2017**, *33*, 258–264.
126. Madrini, B.; Shibusawa, S.; Kojima, Y.; Hosaka, S. Effect of natural zeolite (clinoptilolite) on ammonia emissions of leftover food-rice hulls composting at the initial stage of the thermophilic process. *J. Agric. Meteorol.* **2016**, *72*, 12–19. [CrossRef]
127. Villaseñor, J.; Rodríguez, L.; Fernández, F.J. Composting domestic sewage sludge with natural zeolites in a rotary drum reactor. *Bioresour. Technol.* **2011**, *102*, 1447–1454. [CrossRef]
128. Turan, N.G. Nitrogen availability in composted poultry litter using natural amendments. *Waste Manag. Res.* **2009**, *27*, 19–24. [CrossRef] [PubMed]
129. Hong, J.H.; Park, K.J. Compost biofiltration of ammonia gas from bin composting. *Bioresour. Technol.* **2005**, *96*, 741–745. [CrossRef]

130. Clemens, J.; Cuhls, C. Greenhouse gas emissions from mechanical and biological waste treatment of municipal waste. *Environ. Technol.* **2003**, *24*, 745–754. [CrossRef]
131. Pagans, E.; Font, X.; Sánchez, A. Biofiltration for ammonia removal from composting exhaust gases. *Chem. Eng. J.* **2005**, *113*, 105–110. [CrossRef]
132. Li, Q.; Wang, X.C.; Zhang, H.H.; Shi, H.L.; Hu, T.; Ngo, H.H. Characteristics of nitrogen transformation and microbial community in an aerobic composting reactor under two typical temperatures. *Bioresour. Technol.* **2013**, *137*, 270–277. [CrossRef]
133. Liu, W.; Huo, R.; Xu, J.; Liang, S.; Li, J.; Zhao, T.; Wang, S. Effects of biochar on nitrogen transformation and heavy metals in sludge composting. *Bioresour. Technol.* **2017**, *235*, 43–49. [CrossRef]
134. Awasthi, M.K.; Wang, Q.; Huang, H.; Li, R.; Shen, F.; Lahori, A.H.; Wang, P.; Guo, D.; Guo, Z.; Jiang, S.; et al. Effect of biochar amendment on greenhouse gas emission and bio-availability of heavy metals during sewage sludge co-composting. *J. Clean. Prod.* **2016**, *135*, 829–835. [CrossRef]
135. Malińska, K.; Zabochnicka-Świątek, M.; Dach, J. Effects of biochar amendment on ammonia emission during composting of sewage sludge. *Ecol. Eng.* **2014**, *71*, 474–478. [CrossRef]
136. Agyarko-Mintah, E.; Cowie, A. Biochar increases nitrogen retention and lowers greenhouse gas emissions when added to composting poultry litter. *Waste Manag.* **2017**, *61*, 138–149. [CrossRef]
137. Sun, D.; Lan, Y.; Xu, E.G.; Meng, J.; Chen, W. Biochar as a novel niche for culturing microbial communities in composting. *Waste Manag.* **2016**, *54*, 93–100. [CrossRef]
138. Khan, N.; Clark, I.; Sánchez-Monedero, M.A.; Shea, S.; Meier, S.; Bolan, N. Maturity indices in co-composting of chicken manure and sawdust with biochar. *Bioresour. Technol.* **2014**, *168*, 245–251. [CrossRef] [PubMed]
139. Chen, W.; Liao, X.; Wu, Y.; Liang, J.B.; Mi, J.; Huang, J.; Zhang, H.; Wu, Y.; Qiao, Z.; Li, X.; et al. Effects of different types of biochar on methane and ammonia mitigation during layer manure composting. *Waste Manag.* **2017**, *61*, 506–515. [CrossRef] [PubMed]
140. Zhou, S.; Wen, X.; Cao, Z.; Cheng, R.; Qian, Y.; Mi, J.; Wang, Y.; Liao, X.; Ma, B.; Zou, Y.; et al. Modified cornstalk biochar can reduce ammonia emissions from compost by increasing the number of ammonia-oxidizing bacteria and decreasing urease activity. *Bioresour. Technol.* **2021**, *319*, 124120. [CrossRef]
141. Jia, X.; Wang, M.; Yuan, W.; Shah, S.; Shi, W.; Meng, X.; Ju, X.; Yang, B. $N_2O$ Emission and Nitrogen Transformation in Chicken Manure and Biochar Co-Composting. *Trans. ASABE* **2016**, *59*, 1277–1283. [CrossRef]
142. Guo, H.; Gu, J.; Wang, X.; Yu, J.; Nasir, M.; Zhang, K.; Sun, W. Microbial driven reduction of $N_2O$ and $NH_3$ emissions during composting: Effects of bamboo charcoal and bamboo vinegar. *J. Hazard. Mater.* **2019**, *390*, 121292. [CrossRef]
143. Yang, Y.; Awasthi, M.K.; Wu, L.; Yan, Y.; Lv, J. Microbial driving mechanism of biochar and bean dregs on $NH_3$ and $N_2O$ emissions during composting. *Bioresour. Technol.* **2020**, *315*, 123829. [CrossRef]

processes

MDPI

*Review*

# A Review of Composting Process Models of Organic Solid Waste with a Focus on the Fates of C, N, P, and K

**Zheng Yang [1,*], Furqan Muhayodin [1], Oliver Christopher Larsen [1], Hong Miao [2], Bing Xue [1] and Vera Susanne Rotter [1,*]**

[1] Chair of Circular Economy and Recycling Technology, Technische Universität Berlin, Straße des 17. Juni 135, 10623 Berlin, Germany; furqan.muhayodin@campus.tu-berlin.de (F.M.); oliver.larsen@tu-berlin.de (O.C.L.); bing.xue@tu-berlin.de (B.X.)

[2] College of Mechanical Engineering, Yangzhou University, Yangzhou 225127, China; mh0514@163.com

* Correspondence: zheng.yang@campus.tu-berlin.de (Z.Y.); vera.rotter@tu-berlin.de (V.S.R.)

**Abstract:** To foster a circular economy in line with compost quality assessment, a deep understanding of the fates of nutrients and carbon in the composting process is essential to achieve the co-benefits of value-added and environmentally friendly objectives. This paper is a review aiming to fill in the knowledge gap about the composting process. Firstly, a systematic screening search and a descriptive analysis were conducted on composting models involving the fates of Carbon (C), Nitrogen (N), Phosphorus (P) and Potassium (K) over the past decade, followed by the development of a checklist to define the gap between the existing models and target models. A review of 22 models in total led to the results that the mainstream models involved the fates of C and N, while only a few models involved P and K as target variables. Most of the models described the laboratory-scale composting process. Mechanism-derived models were relatively complex; however, the application of the fractionation of substrates could contribute to reducing the complexity. Alternatively, data-driven models can help us obtain more accurate predictions and involve the fates of more nutrients, depending on the data volume. Finally, the perspective of developing composting models for the fates of C, N, P, and K was proposed.

**Keywords:** composting; organic solid waste; models; nutrients; modeling scale; checklist

**Citation:** Yang, Z.; Muhayodin, F.; Larsen, O.C.; Miao, H.; Xue, B.; Rotter, V.S. A Review of Composting Process Models of Organic Solid Waste with a Focus on the Fates of C, N, P, and K. *Processes* **2021**, *9*, 473. https://doi.org/10.3390/pr9030473

Academic Editor: Antoni Sánchez

Received: 16 February 2021
Accepted: 3 March 2021
Published: 6 March 2021



## 1. Introduction

Organic solid waste (OSW), the solid waste containing organic matters (i.e., food waste, livestock manure, green waste), has been a critical issue for sustainable development due to its continuous increasement in amount and non-recycled treatment [1–3]. Till today, most OSW is still disposed of in unsustainable and conventional ways, such as landfilling and incineration [4], which result in the emission of greenhouse gases and leachate containing heavy metals [5], toxic gases such as sulfur dioxide, dust, heavy metal fumes, and incombustible hydrocarbons, and losses of valuable nutrients [4]. Therefore, the effective management strategies of OSW, including composting, are attached with more importance by relevant stakeholders and policy makers, with the aim of overcoming the challenge of environmental protection, promoting the circular economy and, hence, achieving sustainable development [6–8].

Compared with landfilling and incineration, composting is now one of the most popular technologies to recycle nutrients from organic waste [9], which can significantly shorten the processing cycle and more efficiently recover the nutrients from organic waste [4,10,11]. In China, about 76% of the poultry and livestock manure collected by intensive farming was processed through composting in 2015 [12], which can promote the organic fertilizer production industry and increase the circulation of regional nutrients [13]. Even though fruitful results have been achieved in the research on composing, there remains a large challenge when microorganisms convert complex substrates into ultimately useful products in the composting process, in which some by-products, such as Ammonia ($NH_3$), Carbon dioxide ($CO_2$), Methane ($CH_4$), Nitrous oxide ($N_2O$), etc., are produced to burden the atmosphere [14,15]. The accumulation of P in surface soil can lead to the transfer of Phosphorus (P) to groundwater, which becomes an environmental concern during the compost application [16]. During the composting process, the Carbon (C) loss to the atmosphere ranges from 30% to 63% [17], and the Nitrogen (N) loss ranges from 19% to 42% mainly because of the vigorous $NH_3$ volatilizations, while the Phosphorus (P) loss is less than 2% mostly due to the runoff [18,19]. These data may be different due to the origin of various raw materials. The loss and dissipation of nutrients may not only lead to potential environmental risks, but also reduce the agronomic quality of the composted product [20]. Instead, applying more remaining C from composted fertilizer to the soil can reduce greenhouse gas emissions and sustainably mitigate climate change through storage or sequestration strategies [21]. It will also contribute to the efficiency of other fertilizers by altering soil properties, so as to bring environmental and agronomic co-benefits [22]. Therefore, for composting technology, it is significant to minimize both C and nutrient losses for the production of stable products with high quality.

Generally, the motivation of modeling is to develop mathematical tools to integrate the knowledge with the phenomena, determine the direction of experimental design, evaluate experimental results, test hypotheses, reveal relationships between variables, predict the system development, and design the process and management strategies [23]. Since 1976, mathematical models of composting technology have appeared in the literature [24]. In recent years, many models have been developed to contribute to predicting the distribution of temperature, humidity, solids, oxygen content, and carbon dioxide during the composting process [25–31]. However, from an environmental and agronomic point of view, the focus should be placed on regional C and nutrients for a better understanding of composting technology and assessment of the effectiveness of this sustainable solution [32]. Moreover, the methodology for regional assessments, such as life cycle assessment and material flow analysis, requires the accuracy of the model and a certain number of target variables to be simulated when it is used to simulate and evaluate composting technologies on a regional scale with high accuracy [33]. According to the research of Lauwers et al., the models can be grouped as mechanism-derived models that are established based on the biochemical reaction to reveal more mechanisms and data-driven models focusing more on the experimental data than the process of intermediate reaction [34]. According to the research results from the database of the *Web of Science Core Collection*, the number of papers on the composting process has shown an increase from 74 in 2011 to 114 in 2020. Initially, the focus of relevant research was mechanism-derived models [24], while in recent years, data-driven models based on various algorithms have gained more popularity [35].

Previous articles on the review of composting models usually focused on composting kinetics to discuss the process parameters, such as temperature, water content, pH, and carbon-to-nitrogen ratio (C/N). For instance, Mason reviewed and extensively analyzed composting models proposed in published papers before the end of 2003 [24]. He systematically described the establishment and improvement of the models on heat balance and mass balance during the composting process. Walling et al. conducted a comprehensive review on composting models published in the last 40 years to determine the trend of the composting models in terms of the goal and method, focusing on the research development of composting kinetics, heat balance, and mass (mainly water and oxygen) balance [35]. In recent years, more importance has been given to the simulation of the fates of C, N, P and Potassium (K) in the composting model. However, due to the complexity of the composting process, only a few papers have been published about the systematic review of the modeling of the fates of C and nutrients in the composting process. So, a further study with the application of models is necessary to delve into the fates of C, N, P and K during the composting process. Therefore, the following two research questions are to be addressed with the aim of attaining a deeper understanding and new knowledge based on available studies through the systematic review:

1. What are the key features of existing composting models that involve the fates of C, N, P, and K? (RQ1);
2. How could the gaps between the existing model and the target model be well defined and presented? (RQ2).

The following parts of this paper are structured as follows: Section 2 presents the applied methods to show the process of a systematic review with a descriptive analysis; Section 3 includes the results; Section 4 proposes the guiding perspective of composting models involving the fates of C, N, P, and K, as well as the discussion on the implications of the study, and includes the explanation of how the fate of C, N, P and K in composting can be effectively described through modeling.

## 2. Methods

### 2.1. Literature Screening

A systematic screening search of relevant literature was conducted based on the core collection in the database of *Web of Science* (https://www.webofknowledge.com (accessed on 11 June 2020)), which is considered to cover papers of high quality and in sufficient quantity for a systematic review [36]. The time scope is defined as in the past ten years, from January 2011 to June 2020. The following search rule is used in the advanced search: "(TS = compost) AND (TS = model) AND ((TS = carbon) OR (TS = nitrogen) OR (TS = phosphorus) OR (TS = potassium))", where TS is defined as Topics.

A total of 722 related articles were collected, followed by a precise refining process based on the following three criteria, including: (1) the substrates for composting were OSW; (2) the target variables of modeling objectives involved at least one of C, N, P, and K; (3) the research modeled the process of composting technology. Specifically, the process of study selection and data extraction consists of three steps of results retrieval [37,38], as shown in Figure 1. First, search for articles based on a prioritized search strategy. Then, filter out irrelevant or unsuitable articles according to their titles and abstracts. Third, read the filtered articles in full text. Finally, a total of 22 models were selected for further studies, which are mainly from peer-reviewed journals or conferences such as *Bioresource Technology, Environmental Technology, and Waste Management.*

### 2.2. Data Extraction

In order to further characterize the models, we developed code lists of target variables related to modeling objects, modeling approach types (mechanism-derived model types and data-driven model types), and applied environmental types as indicators to conduct data extraction as shown in Tables S1–S4. From these code lists, we then developed tables shown in Table S5 to describe and summarize the selected models.

### 2.3. Checklist for Model Assessment

A checklist approach was used to define the gap between the reviewed models and the target models. In this study, a checklist was designed according to the target models and the developing process of models. Given the fact that there is no consensus on the best method of evaluating composting models, a brand-new checklist was finally developed and applied here to evaluate the models and help define the gaps of target models, while this method has been applied in other subject areas, such as ecology and medicine [39,40]. The most common questions in the checklists are whether the model clearly describes the objectives of modeling, whether the approach to modeling is reasonable, and whether the sensitivity and accuracy of the model are evaluated [39–41]. Developing a model follows six steps: analyze the problem, formulate a model, solve the model, verify and interpret the model's solution, report by the model, and maintain the model [42]. Furthermore, the emphases in the previous research on composting models, such as the composting substrates [43] and the model's reflection on the mechanism [44], have been combined in developing the checklist to determine three major categories: the start points of the model, the process of modeling, and the internal assessment of models. In addition, to be more in line with our review scope, the target variables of modeling were set on whether the fates of C, N, P and K are all involved as an indicator at the start point of the model. Moreover, since the nutrients' transformation mechanism plays an essential role in studying the balance of elements [44], we set the 7th item to explore the part of the data-driven model of revealing the mechanism in order to better identify the main factors influencing the composting process. The weight of each category is 5-point. As the modeling process of the mechanism-derived models is different from that of the data-driven models, in the second category, different questions were applied to evaluate the two types of models. If we assume the score for the most optimal model is 15, the gap between the model in the checklist and the most optimal one is reflected by 15 subtracting the score for the model. The specific checklist for the composting model is shown in Table 1.

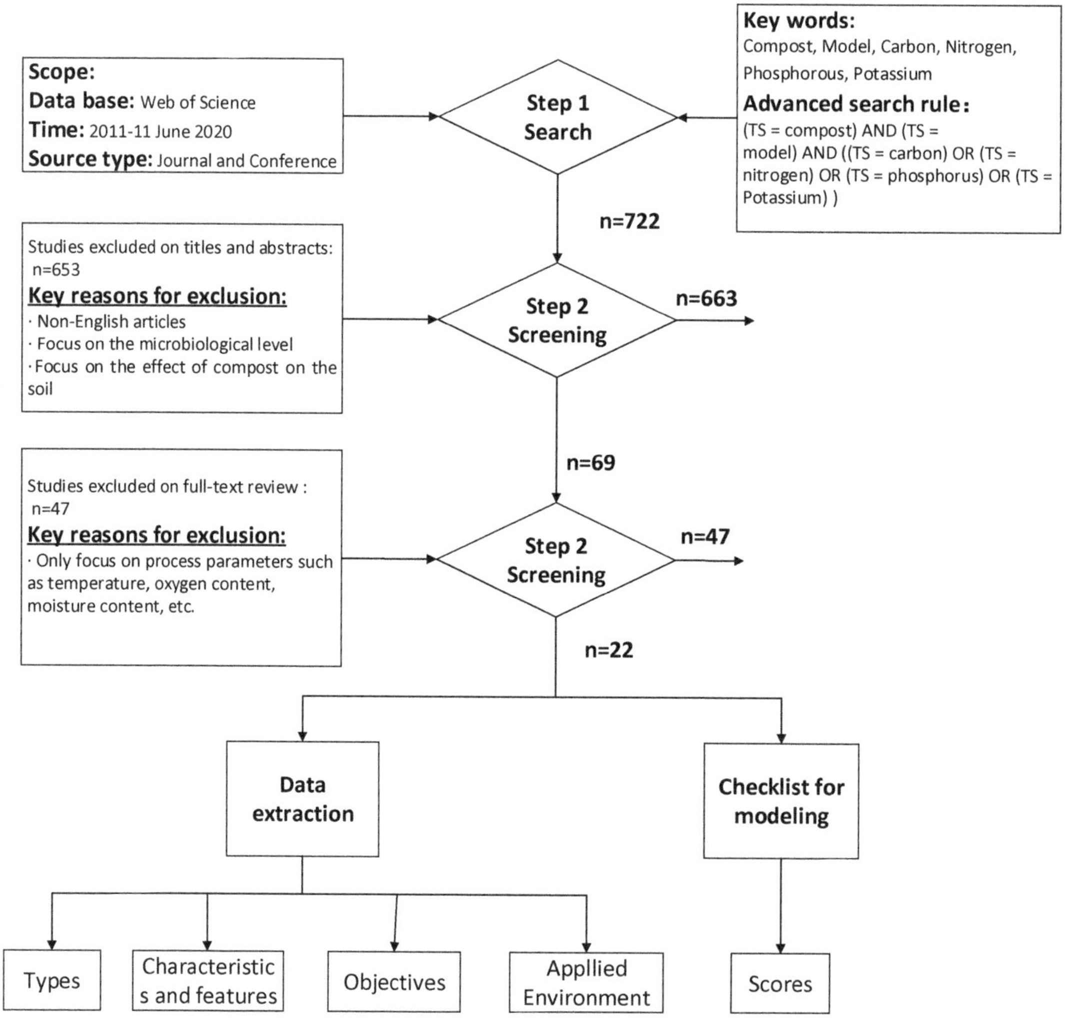


**Figure 1.** Process of study selection and data extraction.

**Table 1.** The checklist for composting models.

| Category | Items | | References |
|---|---|---|---|
| Start points of models | Were the target variables of modeling clearly described? (1 point) | | [39,40] |
| | Do the research objectives fit our review scope (C, N, P, and K)? (3 points) (1 point will be calculated for only one of C, N, P, and K involved in modeling; 2 points will be calculated for 2 or 3 of C, N, P, and K involved in modeling; 3 points will be calculated for all of C, N, P, and K involved in modeling. If partially involved in each related element only, such as $CO_2$ or C/N, 0.5 points will be calculated.) | | |
| | Were the substrates of the study clearly described? (1 point) | | [43,45] |
| Process of modeling | Mechanism-derived models | Data-driven models | |
| | Does the selection equation in the model clearly list the reference basis? (1 point) | Does the study identify the sources of the data and describe how the data were collected clearly? (1 point) | [39,41,42] |
| | Were the assumptions about the model clearly described? (1 point) | Was the modeling approach used clearly described? Does it include the reasons for adopting this approach (1 point) | [39,40,42] |
| | Was the basis for the selection of relevant parameters clearly described? (1 point) | Was the basis for the selection of variables clearly described? (1 point) | [24,40] |
| | How about the complexity of the models? (1 point, 0.5 points, or 0 will be calculated for Not complicated, Complicated, and Very complicated, respectively) | How well does the model reflect the composting process? (1 point, 0.5 points, or 0 will be calculated for Well reflect, Partly reflect, and Not reflect, respectively) | [42,44] |
| | Was the platform/software clearly described to solve/simulate the model? (1 point) | | [42] |
| Internal assessment of models | Was the sensitivity analysis conducted? (1 point) | | [40,46] |
| | Were experiments conducted to compare the models? (1 point) | | [39] |
| | Was the accuracy evaluation method of the models clearly described? (1 point) | | [34,42] |
| | How about the accuracy of the models? (2 points, 1 point, or 0 will be calculated for Very accurate, Relatively accurate, and Not accurate or not mentioned, respectively) | | [42] |

Out of the 12 questions, 9 were judged between yes or no, and the other three were scored based on the reality of the model. With reference to Wijewardhana et al. and Harris et al., we applied multiple reviewers to the checklist to ensure relative objectivity. All indicator questions were rated by three reviewers who have a research background in modeling or composting technology [39,40]. For yes/no questions, a discussion with the author would be proposed in case of a different judgment. For questions that needed to be scored according to circumstances, an average score was calculated. What is worth mentioning is that, in order to make the whole procedure as objective as possible, two rounds of review were conducted on the checklist and results, one internally by the authors and the other by an invited expert from the Institute of Soil Science, Chinese Academy of Sciences, an the external reviewer.

## 3. Results

### 3.1. Overview of Reviewed Models

The substrates, modeling approaches, and target variables of objectives for 22 referred models are shown in Figure 2. The 22 models were divided into two main categories based on the modeling approaches: 10 mechanism-derived models and 12 data-driven models. In particular, semi-empirical models fell in between [44], which are established based on mechanism-derived models but modified with experimental data. Since these three semi-empirical models were developed from a process perspective, they were also summarized in the mechanism-derived model in this section. The composting substrates of these models were mainly related to two categories including municipal solid waste (MSW) and agricultural waste. The target variables involved in the simulation, however, were mostly C and N, and to a lesser extent, P and K.

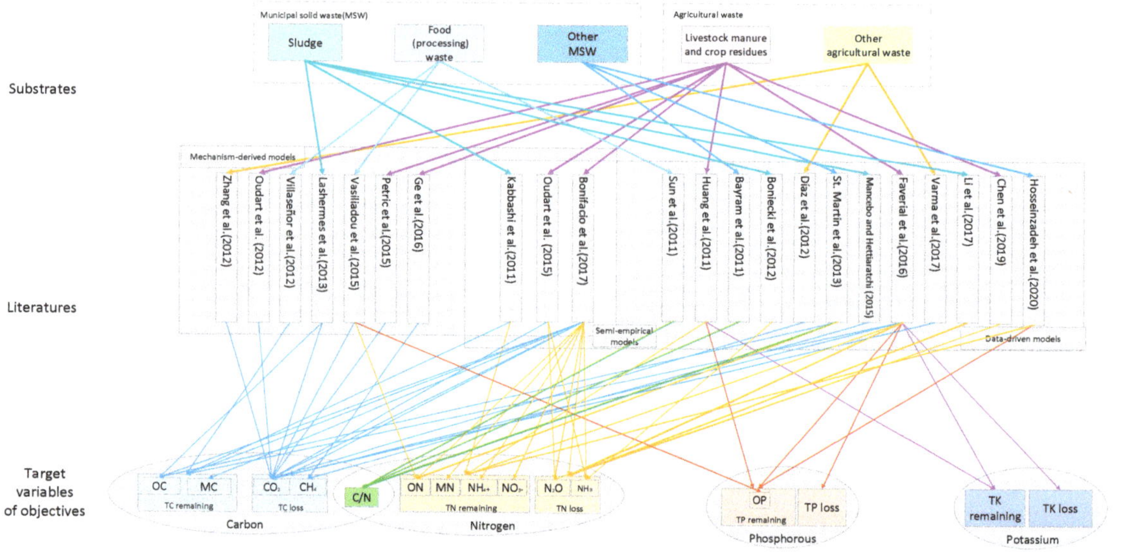

**Figure 2.** An overview of 22 reviewed models. Notes: The abbreviations are defined as follows: **MSW** (municipal solid waste), **OC** (organic carbon), **MC** (microbial biomass carbon), **TC** (total carbon), **ON** (organic nitrogen), **C/N** (carbon-to-nitrogen ratio), **MN** (microbial nitrogen), **OP** (organic potassium), **TP** (total phosphorus), **TK** (total potassium).

### *3.2. Composting Substrates and Target Variables*

The specific substrates of these models involved MSW and agricultural waste, as shown in Figure 3a. MSW mainly includes sludge (n = 5) and food waste or food processing waste (n = 3). In comparison, other types of municipal waste have been studied, including cardboard, boxwood leaves, and sawdust (n = 3). In these models, most of the simulation of agricultural waste concentrated on livestock manure and crop residues, such as pig manure, chicken manure, and cattle manure mixed with straws of rice, wheat, and corn (n = 8). In addition, other types of agricultural wastes include vegetable wastes and fruit leaves (n = 3).

To address the challenges posed by the complexity of the substrates for compost modeling, fractionation of the substrates was applied to separate the organic matter into multiple components. Simply put, the substrates are divided into three categories, namely, soluble, insoluble, and inert substrates [44,47–50]. Furthermore, a more detailed fractionation method was applied, in which the organic matters were divided into five compartments: the easily degradable and soluble; slowly degradable and soluble; hemicelluloses, cellulose, and lignin fractions [51,52]. With this method of fractionation, the degradation process of the organic matters can be described according to different degradation kinetics, thereby improving the accuracy of the model, and at the same time, providing a solution to the modeling of complex substrate composting.

Since the review scope of this paper was the fates of carbon and nutrients, only the target variables related to C, N, P, and K in modeling were included. There were two parts in each element: the remaining and the lost. It can be seen from Figure 3b that most models involved the simulation of C and N. Models involving carbon mainly included organic carbon (OC) (n = 3) and microbial carbon (MC) (n = 1). There was also research on the remaining of total carbon (TC) (n = 4). The simulation of carbon loss mainly involved $CO_2$ (n = 9) and $CH_4$ (n = 2). In terms of nitrogen, Bonifacio et al., St Martin et al. and Vasiliadou et al. developed models for organic nitrogen (ON) (n = 3) [33,49,53]. As for total nitrogen (TN) loss, Li et al. and Faverial et al. modeled this part as a whole variable (n = 2) [15,54]; others focused on the emissions of $N_2O$ (n = 3) and $NH_3$ (n = 3). There were some models related to the C/N that are considered to play a key role in the composting process, and these models also involve the mass balance of C and N (n= 3). Vasiliadou et al., Faverial et al., and Huang et al. have developed models for the mass balance of total phosphorus (TP) (n = 3) [15,49,55]. The research by Faverial et al. and Huang et al. also involved the model of total potassium (TK) (n = 2) [15,55].

In addition, mechanism-derived models mainly simulated the relevant mass balance of C and N, and, to a lesser extent, the mass balance of P. In contrast, the data-driven models could cover a broader range of simulated objects and even involved K. However, there were no mechanism-derived models that included K in the selected research.

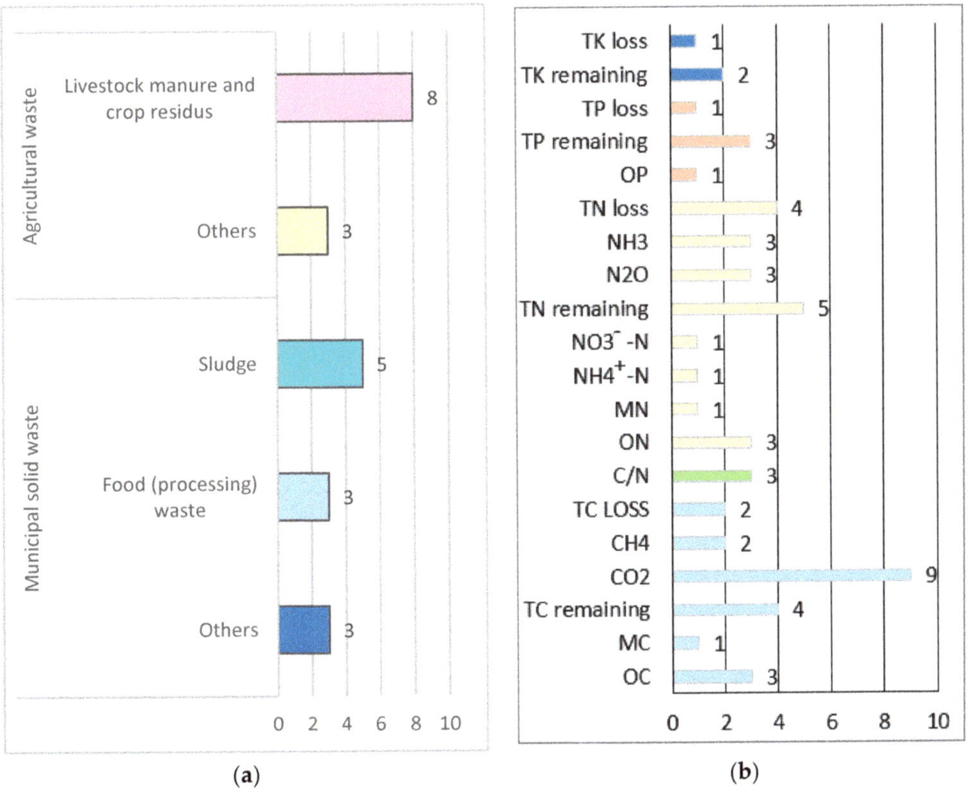


**Figure 3.** The specific situation regarding the (**a**) substrates and (**b**) the target variables of modeling objectives. Notes: The abbreviations are defined as follows: **OC** (organic carbon); **MC** (microbial carbon); **TC** (total carbon); **ON** (organic nitrogen); **C/N** (carbon-to-nitrogen ratio); **MN** (microbial nitrogen); **OP** (organic potassium); **TP** (total phosphorus); **TK** (total potassium).

### 3.3. *Modeling Approaches*

#### 3.3.1. Mechanism-Derived Models

The mechanism-derived models are generally based on mass balance, energy balance, and kinetics [56]. Composting kinetics describes methods of controlling the rate of waste degradation through environmental factors, such as temperature, oxygen utilization, and moisture. So far, various kinetics models for biomass degradation through composting have been developed based on the physical and biochemical characteristics of composting materials [57]. A summary of 10 mechanism-derived models and modeling objectives is shown in Table 2.

**Table 2.** Summary of 10 mechanism-derived models and modeling objectives.

| No. | References | Mechanism-Derived Model Type Involved | Related Modeling Objectives |
|---|---|---|---|
| 1 | Zhang et al., 2012 [51] | Monod kinetics model<br>First-order kinetics model<br>Mass balance model | $CO_2$ corresponding to mineralization<br>(% of initial total organic carbon) |
| 2 | Oudart et al., 2012 [47] | | $CO_2$ emission rate |
| 3 | Lashermes et al., 2013 [52] | | OC and $CO_2$ corresponding to mineralization<br>(% of initial total OC) |
| 4 | Villaseñor et al., 2012 [50] | First-order kinetics model | C degradation<br>(% of DM) |
| 5 | Vasiliadou et al., 2015 [49] | Monod kinetics model<br>First-order kinetics model<br>Mass balance model<br>Heat (energy) balance model | Insoluble organic matter mass, insoluble N and P mass, and $CO_2$ emission volume |
| 6 | Petric and Mustafić 2015 [56] | Monod kinetic model<br>Mass balance model<br>Heat (energy) balance model | $CO_2$ mass |
| 7 | Ge et al., 2016 [48] | First-order kinetics model<br>Michaelis–Menten kinetics model<br>Energy balance model<br>Mass balance model | $CH_4$ emission rate |
| 8 | Kabbashi 2011 [58] | Semi-empirical model<br>Multi-stage model | The remaining of TC and TN<br>(% of DM) |
| 9 | Oudart et al., 2015 [44] | Semi-empirical model<br>Process-based model | Production yield of $CO_2$, $N_2O$ and $NH_3$ |
| 10 | Bonifacio et al., 2017 [33,59] | | OC, MC, ON, MN, $NH_4^+$, $NO_3^-$<br>(% of DM),<br>and emission rates of $CO_2$, $N_2O$ and $NH_3$ |

**OC** (organic carbon); **TC** (total carbon); **TN** (total nitrogen); **MC** (microbial carbon); **ON** (organic nitrogen); **MN** (microbial nitrogen); **DM** (dry matter).

The common kinetics model is the first-order kinetics model (n = 6) related to the degradation of volatile solids or the utilization of oxygen. Hence, it has a close connection with the fate of C. The first-order kinetics model is based on temperature, oxygen, moisture, biodegradable volatile solids (BVS), and free space as parameters that affect the rate of degradation [60,61].

Another widely used kinetics model is the Monod kinetics model (n = 5), which was developed from the mechanical or deductive point of view by integrating the basic principles of physics, chemistry, and microbiology involved in the composting process [56,62,63].

The derivation of each kinetics model focuses on their mathematical formulas, which allows them to explain certain processes in composting. In the first-order kinetics model, the substrate concentration is used as the primary force determining the reaction rate, while the Monod kinetics model involves microbial activity, which makes the model more realistic.

Semi-empirical models are based on the mechanism with test data to modify and determine their model parameters. This approach is different from other mechanism-derived models, which requires a comprehensive understanding of the process. Unlike data-driven models that rely on large amounts of data, it is developed based on internal processes or stages. Oudart et al. simulated the interaction of nitrogen and carbon during animal manure composting based on the main processes governing carbon and nitrogen transformations [44]. Then, models were analyzed and simulated according to the experimental data. Bonifacio et al. developed a process-based model for simulating cattle manure compost windrows [33,64]. In their research, the fate of C and N through processes affected by compost windrows was established. Combined with a large amount of empirical data, the parameters were determined to study the mass balance of C and N.

In order to describe more variables, more equations and parameters are required, leading to the complexity of models. In mechanistically derived models, the studies by Bonifacio et al. and Oudart et al. involved more related modeling objectives [33,44]. The former included 10 equations and 52 parameters, while the latter included more, with 26 equations and nearly 90 parameters. In addition to using mathematical models to simulate microbial growth, nitrification, denitrification and other biochemical process reactions, some physical processes were also described. For example, Bonifacio et al. incorporated the leaching and runoff of $NO_3^-$ as well as ammonia volatilization, into the model [33]. Oudart et al. also considered ammonia volatilization [44].

### 3.3.2. Data-Driven Models

Data-driven models are usually accompanied by experimental and empirical data collection to ensure the effective prediction of fundamental parameters [34], thereby establishing a reliable relationship between the model and the prediction of essential parameters or variables. A summary of 12 data-driven models and simulation objects is shown in Table 3.

**Table 3.** The summary of 12 data-driven models and simulation objects.

| No. | References | Modeling Type | Input Variables | Target Variables Related to Modeling Objects |
|---|---|---|---|---|
| 1 | Sun et al., 2011 [65] | Genetic algorithm aided by the stepwise cluster analysis method | $NH_4^+ - N$ concentration, moisture content, ash content, mean temperature, and mesophilic bacteria biomass | C/N |
| 2 | Huang et al., 2011 [55] | Linear regression analysis | pH, EC, and DM content | The remaining TN, TP, and TK (% of DM) |
| 3 | Bayram et al., 2011 [66] | ANN model<br>MLR model | Food and yard percentage, ash and scoria percentage, moisture content, fixed carbon content, the total proportion of organic matter, high, calorific value, and pH | C/N |
| 4 | Hosseinzadeh et al., 2020 [67] | ANN model<br>MLR model | pH, EC, C/N, $NH_4^+/NO_3^-$, water-soluble carbon, dehydrogenase enzyme, and total phosphorus | The remaining TN and TP (% of DM) |
| 5 | Boniecki et al., 2012 [59] | ANN model | Time, temperature, pH, EC, DM concentration, C/N, $NH_4^+ - N$ concentration | $NH_3$ emissions (% of air released from bioreactor chamber) |
| 6 | Díaz et al., 2012 [68] | An adaptive network-based fuzzy inference system | Aeration rate, moisture content, particle size, and time | $CO_2$ emission rate |
| 7 | St Martin et al., 2014 [53] | Critical exponential function<br>Rectangular hyperbola function<br>(Double) Fourier function<br>MLR model | Composting formula, time and composting formula interacting through time | TOC and TKN (% of DM) |

**Table 3.** *Cont.*

| No. | References | Modeling Type | Input Variables | Target Variables Related to Modeling Objects |
|---|---|---|---|---|
| 8 | Faverial et al., 2016 [15] | Bayesian network model | Total C, N, lignin, P and K contents, pH, and loss of mass | The remaining, and loss of, TN, TP, and TK (% of DM) |
| 9 | Mancebo and Hettiaratchi 2015 [69] | Regression model | Air-filled porosity, moisture content, and dissolved OC content | $CH_4$ emission rate |
| 10 | Li et al., 2017 [54] | | Sucrose-adding ratio, adding time, sucrose concentration | The loss TN ration |
| 11 | Varma et al., 2017 [70] | RBF neural network model | Moisture content, pH, EC, TOC, TKN, soluble biochemical oxygen demand, $NH_4^+ - N$ concentration, available phosphorous, C/N, total phosphorous, oxygen uptake rate, Na, K, Ca | $CO_2$ emission rate |
| 12 | Chen et al., 2019 [71] | Backpropagation neural network model Linear regression model | Moisture content, C/N, aeration rate, and superphosphate content | Proportion of $N_2O$ on TN |

**ANN** (artificial neural network); **BP** (backpropagation); **RBF** (radial basis functional); **MLR** (multiple linear regression); **EC** (electrical conductivity); **DM** (dry matter); **C/N** (carbon-to-nitrogen ratio); **TN** (total nitrogen); **TP** (total phosphorus); **TK** (total potassium); **TOC** (total organic carbon); **TKN** (total Kjeldahl nitrogen).

Artificial neural network (ANN) is most widely used in data-driven models (n = 6), which is designed to simulate the biological nervous system's response to real-world tasks [72]. In the reviewed articles, different types of neural networks have been studied, including multilayer perceptron (MLP) [59,67], backpropagation (BP) [71], and radial basis functional (RBF) [70]. BP is a systematic approach to training MLP. Bayram et al. (2011) used the MLP trained with the BP algorithm to develop models for simulating C/N of MSW composting [66].

Linear regression analysis of data is a monitoring technique used to model target values based on independent predictors [72]. The composting process can be modeled based on one variable (single regression) model or multiple variables (multiple linear regression (MLR)) model. St Martin et al. used different function models to simulate different parameters of the composting process, leading to the recognition that the composting temperature and OC are best described by the critical exponential function and the rectangular hyperbolic function, respectively [53]. ON, C/N, and pH are best described by double Fourier functions, while electrical conductivity (EC) is best described via Fourier functions. Huang et al. discussed the efficiency and feasibility of nutrient elements in chicken manure during composting with physical and chemical properties, such as pH, EC and DM [55]. It can be concluded that DM is a better predictor constructed as a single linear regression of nutrients, while DM and pH are more notable for MLR. Since MLR also involves multiple variables, it is usually compared with the ANN model in articles (n = 3). However, in terms of accuracy, the ANN model performed better in all three articles. Other models, such as Bayesian network models [15] and Genetic algorithms [65], are all used in data-driven models.

The selection of input is an important step in developing the data-driven model. As can be seen in Table 3, pH is the most commonly used input variable (n = 7), which has a great influence on the decay, odor emission, nutrient conversion, and loss rate in the composting process [15,59]. Others, such as moisture content (n = 6), EC (n = 5), C/N (n = 4), and temperature (n = 3), are also commonly used as input variables.

### 3.4. Application Scales

Overall, as can be seen from Table 4, most of the mathematical models are still in the scope of the laboratory (n = 18). Bonifacio et al. and Oudart et al. developed semi-empirical models for the farm scale since the simulation and data collection were based on a farm over several years [33,44,64]. Huang et al. modeled based on data from composting plants in the perspective of a factory [55]. In addition, Vasiliadou et al. conducted a modeling study in the scale of the olive plant from the industrial plant scale [49]. According to the modeling approaches, both mechanism-derived and data-driven models could be studied at different scales. The research on the lab scale is more concerned with the composting reaction process itself through describing the target variables in detail. In contrast, research from the industrial plant scale and farm scale tends to account for more indicators.

**Table 4.** The numbers of reviewed models according to applied scales.

| Applied Scales | Number of Reviewed Models | |
|---|---|---|
| | Mechanism-Derived Models | Data-Driven Models |
| Lab scale | 7 | 11 |
| Industrial plant scale | 1 | 1 |
| Farm scale | 2 | 0 |

### 3.5. Sensitivity Analysis and Validation

Sensitivity analysis and model validation are the main approaches to evaluating models [42]. Since the mechanism-derived models have more parameters, sensitivity analysis on the model is often conducted to assess the uncertainty of model parameters (n = 6). It was noted in these studies that the maximum growth rate coefficient [49,51,56] and mortality constant have a more considerable influence on the composting process parameters [51,52]. For the data-driven model, in addition to the conventional sensitivity analysis (n = 7), there is the adopting analysis of variance (ANOVA), which can also be used to achieve the purpose of sensitivity analysis (n = 3) in terms of selected input variables. For instance, the ANOVA of St Martin et al. indicated that composting formula, time and composting formula interacting through time had a significant impact on the variables such as temperature, total organic carbon (TOC), total Kjeldahl nitrogen (TKN), C/N, pH, and EC [53]. Li et al. showed that the effect of addition ratio and addition time on nitrogen loss was statically significant at the 95% confidential level through ANOVA [54].

After obtaining a model, to verify the accuracy of the model, the determination coefficient ($R^2$) (n = 12) and root-mean-square error (RMSE) (n = 6) are the most commonly used methods to evaluate the quality of the fitting accuracy under the assumption that the parameters of the model are normally distributed. The calculation formulas are as follows [47]:

$$\mathrm{RMSE} = \frac{100}{E} \cdot \sqrt{\sum_{i=1}^{n} (S_i - E_i)^2 / n} \quad (1)$$

$$\mathrm{R}^2 = \frac{\sum_1^n (S_i - E)}{\sum_1^n (E_i - E)} \quad (2)$$

where $E$, $S_i$, $E_i$ and $n$ are referred to as the averages of experimental values, simulated values, experimental values, and the number of samples, respectively.

Others, such as Nash–Sutcliffe efficiency (NSE), a normalized statistic used to determine the relative size of the residual variance compared to the variation of the measured data, is also used to evaluate a model's quality [51,52,70]. St Martin et al. adopted a parallel curve analysis to carry out variance accumulation analysis of the effect of compost type and time on physical and chemical parameter models [53].

### 3.6. Gaps with the Target Models Reflected by the Checklist

With the checklist, the scores of gaps ranged from 1.3 to 7.7, which can be seen in Figure 4. The model's scores were only obtained in the checklist that we created to show the gaps between the target models. The checklist could efficiently describe the fates of C, N, P, and K during composting. It was not aimed to completely distinguish the advantages and disadvantages of models, but largely focused on checking whether these models fit the scope and subject of the review, and how well they fitted the procedures modeled. It can be seen from Figure 4 that the research of Faverial et al. was more in line with the scope of the review, while the overall modeling was also in line with the specification, having an excellent performance in accuracy [15]. The paper of Chen et al., a conference paper with limited space, also attracted our attention, in which their scores were affected as some modeling procedures may not be described in details [71]. The starting points of the model involve the target variables of modeling objectives; however, there are many models that do not fully include C, N, P, and K. When the starting points of the model are excluded from checklist results, there are more models that also perform very well.

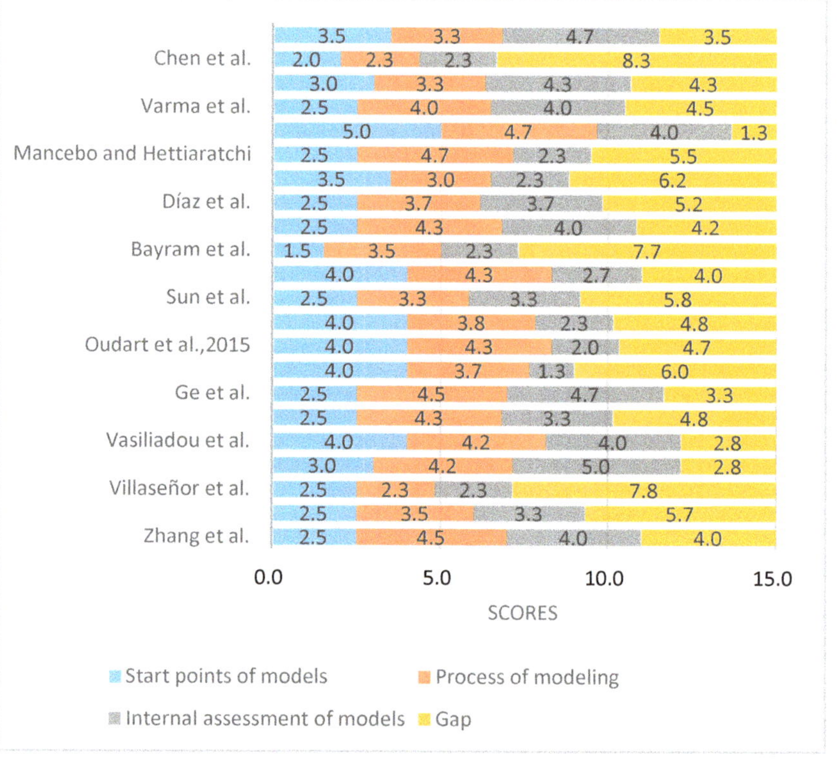


**Figure 4.** Results of the checklist for defining gaps.

## 4. Discussion

The purpose of developing models for the fates of C, N, P, and K is to improve process operations and, more importantly, deepen our understanding of the process, so as to improve the utilization of nutrients and reduce greenhouse gas emissions to achieve co-benefits for building the regional circular economy [73]. Therefore, the mechanics and the accuracy of the models are significant for the realization of the above purpose. Mechanism-derived models are ideal models for revealing mechanisms; however, a lot of effort is required due to the complexity of the models. Moreover, the composting process is a biochemical reaction process that involves physical phenomena, such as volatilization and leaching [74], which are often ignored by most of the mechanism-derived models, resulting in compromised accuracy. With the study of microbial communities, more and more composition information about a data-rich microbial community will be gained to significantly improve the performance of the model. For example, further knowledge of microbial growth coefficients and mortality coefficients, etc., contributed a lot to the description of microbial activity in the composting process [34]. Additionally, in order to be able to simulate more nutrients, such as P and K, a focus on this part of the research would advance the development of mechanism-derived models of composting that involve more fates of nutrients. As Oudart et al. mentioned, black-box models such as data-driven models, due to the ignorance of complex reaction processes, often cause difficulty in explaining the differences between the results of simulation and observation [44]. The selection of input variables, sensitivity and uncertainty analysis is precisely the part that can react to the mechanism of the composting process. So, for data-driven models, this study will advance their role in revealing mechanisms. The issue of data reliability, however, has always been one of the top priorities for data-driven models. The application of advanced monitoring technology in the composting process will provide the model with certain intermediate process parameters, thereby reducing the possible errors.

At present, most of the models are at lab scale, which tend to focus on the fate of the C and nutrients in the process during composting. For the models on industrial plant or farm scales, more factors will be incorporated, such as N run off and leaching on the surface [33], as the data come from a wider perspective. In order to describe the modeling of composting in agricultural production activities on a regional scale, more indicators should be included, such as greenhouse gas emissions, nutrient losses, and proxies for ecosystem service that result from material exchanges among stakeholders [75].

Meanwhile, the development of open science will also promote the progress of the model. It is worth mentioning that among the 22 selected models, the model of Bonifacio et al. is based on the Integrated Farm System Model (IFSM) [33], which is a public integrated farm research tool for many physical and biological processes [76]. In addition, huge amounts of empirical data are included to provide support for the development of the model. In addition, it can be found that the researchers working on these models gradually began to pay attention to the significance of open science for scientific progress. For instance, Faverial et al. obtained the highest score in the checklist and their paper can be openly accessed [15]. Another treatment technology, anaerobic digestion (AD), a unified and open model of Anaerobic Digestion Model No. 1 (ADM1) was proposed as early as in 2002, which undoubtedly has played a positive role in the development of the AD models. Furthermore, some databases such as PHYLLIS 2 database are gradually being established, which provide a large amount of reliable, high-quality, and shared biomass processing data as strong support for the development of data-driven models.

Regarding this research, some limitations are also worth our attention: First, the research on latest models involving the fates of C, N, P, and K was conducted in the time scope of past decade, and only English-written papers from Web of Science were selected, which means less involved models were selected. Second, as we focused more on C and nutrients balance, the overview of composting modeling in our research is not as comprehensive as that in some other review papers regarding modeling of the composting process [35]. In fact, as was mentioned by Mason and Walling et al., heat balance, moisture content balance, and oxygen content balance have an essential impact on composting. Furthermore, there is inevitable subjectivity when the checklist is used to assess models [24,35]. These models and scientific articles are peer-reviewed and have a high level of creativity. However, data extraction through listing codes and the checklist evaluation method we applied are based on our review scope and more in line with the modeling procedure. Therefore, a degree of subjectivity may occur in our research of the checklist, mainly due to the professional background of the reviewers. More reviewers or multiple rounds of reviews would help to reduce the subjectivity. More importantly, our study intends to provide guidance for future model development in the field of modeling on the fates of C, N, P, and K during composting process.

**5. Conclusions**

In this study, a systematic review was performed on the composting models involving C, N, P, and K. After reviewing the existing literature, 22 composting models were selected with the process of study selection. The application of a code-listing data extraction method could provide a framework for a better summary and cross-model comparisons. In addition, the characteristics and features of these 22 models were presented after data extraction. A checklist for composting models was created to define the gap between existing models and target models. The aim was to find the best fitting model for the composting of various types of substrates. According to the modeling approaches, 22 models were divided into two categories: the mechanism-derived models and the data-driven models. The results of the checklist showed that the score of the mechanism-driven models was slightly higher than that of the data-driven models. The main reason is that the description of the selection basis of variables is ignored in some data-driven models, resulting in a deficiency in highlighting the mechanism of the composting process.

The mechanism-derived model does not involve the simulation of the mass balance of K. Through the sensitivity analysis in these studies, it is found that maximum growth rate coefficients and mortality constants are the main factors for the kinetics parameters. Although the mechanism-derived model is complicated, adopting the method of substrates fractionation has reduced the complexity and improved the accuracy. At the same time, proposing a model framework such as ADM1 is also an approach to reducing the complexity of the model. With the development of artificial intelligence algorithms, data-driven models can cover more target variables involving more nutrients. However, how to reveal the mechanism of the composting process based on the selection of input variables and the establishment of a reliable database still needs some further research.

From the perspective of the model supporting the circular economy assessment at a regional scale, the focus should be on more indicators and high accuracy of models. On a larger scale, more indicators will be included in the modeling to allow for a more comprehensive assessment of circularity. At the same time, it is a scale-up process that requires a high level of accuracy for small scale models in order to ensure the accuracy of the regional model. These set requirements for the future development of composting models.

**Supplementary Materials:** The following are available online at https://www.mdpi.com/2227-9717/9/3/473/s1, Table S1: Code list of target variables related to modeling objects, Table S2: Code list of mechanism-derived model types, Table S3: Code list of data-driven model types, Table S4: Code list of applied scale types, Table S5: Summary of 22 models.

**Author Contributions:** Conceptualization and supervision, V.S.R.; methodology, Z.Y.; formal analysis, Z.Y., F.M., H.M., O.C.L., and B.X.; data curation, Z.Y.; writing—original draft preparation, Z.Y.; writing—review and editing, Z.Y., V.S.R., and B.X. All authors have read and agreed to the published version of the manuscript.

**Funding:** This study was supported by the China Scholarship Council Scholarship (No. 201908320362) and the project "URA" (No. 01LE1804A1) funded by German Federal Ministry of Education and Research and German Aerospace Center.

**Acknowledgments:** The authors would also like to thank Jiawei Hu for his contribution to the checklist scoring work and Bo Sun of the Institute of Soil Science, Chinese Academy of Sciences, for reviewing and suggesting the checklist results. We acknowledge support by the German Research Foundation and the Open Access Publication Fund of Technische Universität Berlin.

**Conflicts of Interest:** The authors declare no conflict of interest.

## Abbreviations

| | |
|---|---|
| AD | Anaerobic digestion |
| ADM1 | Anaerobic Digestion Model No. 1 |
| ANN | Artificial neural network |
| ANOVA | Adopting analysis of variance |
| BP | Backpropagation |
| BVS | Biodegradable volatile solids |
| C | Carbon |
| $CH_4$ | Methane |
| $CO_2$ | Carbon dioxide |
| C/N | Carbon-to-nitrogen ratio |
| DM | Dry matter |
| EC | Electrical conductivity |
| IFSM | Integrated Farm System Model |
| K | Potassium |
| MC | Microbial carbon |
| MLP | Multilayer perceptron |
| MLR | Multiple linear regression |
| MN | Microbial nitrogen |
| MSW | Municipal solid waste |
| N | Nitrogen |
| $NH_3$ | Ammonia |
| $N_2O$ | Nitrous oxide |
| NSE | Nash–Sutcliffe efficiency |
| OC | Organic carbon |
| ON | Organic nitrogen |
| P | Phosphorus |
| $R^2$ | Determination coefficient |
| RBF | Radial basis functional |
| RMSE | Root-mean-square error |
| TC | Total carbon |
| TK | Total potassium |
| TKN | Total Kjeldahl nitrogen |
| TN | Total nitrogen |
| TOC | Total organic carbon |
| TP | Total phosphorus |
| VOC | Volatile organic compounds |

## References


1. Yalcinkaya, S. A spatial modeling approach for siting, sizing and economic assessment of centralized biogas plants in organic waste management. *J. Clean. Prod.* **2020**, *255*, 120040. [CrossRef]
2. Dhanya, B.S.; Mishra, A.; Chandel, A.K.; Verma, M.L. Development of sustainable approaches for converting the organic waste to bioenergy. *Sci. Total Environ.* **2020**, *723*, 138109. [CrossRef] [PubMed]
3. Guo, H.-N.; Wu, S.-B.; Tian, Y.-J.; Zhang, J.; Liu, H.-T. Application of machine learning methods for the prediction of organic solid waste treatment and recycling processes: A review. *Bioresour. Technol.* **2021**, *319*, 124114. [CrossRef]
4. Chen, T.; Zhang, S.; Yuan, Z. Adoption of solid organic waste composting products: A critical review. *J. Clean. Prod.* **2020**, *272*, 122712. [CrossRef]
5. Mor, S.; Ravindra, K.; Dahiya, R.P.; Chandra, A. Leachate characterization and assessment of groundwater pollution near municipal solid waste landfill site. *Environ. Monit. Assess.* **2006**, *118*, 435–456. [CrossRef] [PubMed]
6. Onwosi, C.O.; Igbokwe, V.C.; Odimba, J.N.; Eke, I.E.; Nwankwoala, M.O.; Iroh, I.N.; Ezeogu, L.I. Composting technology in waste stabilization: On the methods, challenges and future prospects. *J. Environ. Manag.* **2017**, *190*, 140–157. [CrossRef]
7. Mayer, F.; Bhandari, R.; Gäth, S. Critical review on life cycle assessment of conventional and innovative waste-to-energy technologies. *Sci. Total Environ.* **2019**, *672*, 708–721. [CrossRef] [PubMed]
8. Wainaina, S.; Awasthi, M.K.; Sarsaiya, S.; Chen, H.; Singh, E.; Kumar, A.; Ravindran, B.; Awasthi, S.K.; Liu, T.; Duan, Y.; et al. Resource recovery and circular economy from organic solid waste using aerobic and anaerobic digestion technologies. *Bioresour. Technol.* **2020**, *301*, 122778. [CrossRef] [PubMed]
9. Aguelmous, A.; El Fels, L.; Souabi, S.; Zamama, M.; Hafidi, M. The fate of total petroleum hydrocarbons during oily sludge composting: A critical review. *Rev. Environ. Sci. Biotechnol.* **2019**, *18*, 473–493. [CrossRef]

10. Kulkarni, M.B.; Ghanegaonkar, P.M. Biogas generation from floral waste using different techniques. *Glob. J. Environ. Sci. Manag.* **2019**, *5*, 17–30. [CrossRef]
11. Fernandez-Bayo, J.D.; Yazdani, R.; Simmons, C.W.; VanderGheynst, J.S. Comparison of thermophilic anaerobic and aerobic treatment processes for stabilization of green and food wastes and production of soil amendments. *Waste Manag.* **2018**, *77*, 555–564. [CrossRef]
12. Liu, Z.; Wang, X.; Wang, F.; Bai, Z.; Chadwick, D.; Misselbrook, T.; Ma, L. The progress of composting technologies from static heap to intelligent reactor: Benefits and limitations. *J. Clean. Prod.* **2020**, 122328. [CrossRef]
13. Cerda, A.; Artola, A.; Font, X.; Barrena, R.; Gea, T.; Sánchez, A. Composting of food wastes: Status and challenges. *Bioresour. Technol.* **2018**, *248*, 57–67. [CrossRef]
14. Li, H.; Chen, Z.; Fu, D.; Wang, Y.; Zheng, Y.; Li, Q. Improved ADM1 for modelling C, N, P fates in anaerobic digestion process of pig manure and optimization approaches to biogas production. *Renew. Energy* **2020**, *146*, 2330–2336. [CrossRef]
15. Faverial, J.; Cornet, D.; Paul, J.; Sierra, J. Multivariate Analysis of the Determinants of the End-Product Quality of Manure-Based Composts and Vermicomposts Using Bayesian Network Modelling. *PLoS ONE* **2016**, *11*, e0157884. [CrossRef] [PubMed]
16. Galvez-Sola, L.; Morales, J.; Mayoral, A.M.; Marhuenda-Egea, F.C.; Martinez-Sabater, E.; Perez-Murcia, M.D.; Bustamante, M.A.; Paredes, C.; Moral, R. Estimation of phosphorus content and dynamics during composting: Use of near infrared spectroscopy. *Chemosphere* **2010**, *78*, 13–21. [CrossRef] [PubMed]
17. Tiquia, S.M.; Richard, T.L.; Honeyman, M.S. Carbon, nutrient, and mass loss during composting. *Nutr. Cycl. Agroecosyst.* **2002**, *62*, 15–24. [CrossRef]
18. Eghball, B.; Power, J.F.; Gilley, J.E.; Doran, J.W. Nutrient, Carbon, and Mass Loss during Composting of Beef Cattle Feedlot Manure. *J. Environ. Qual.* **1997**, *26*, 189–193. [CrossRef]
19. Zhao, S.; Schmidt, S.; Qin, W.; Li, J.; Li, G.; Zhang, W. Towards the circular nitrogen economy—A global meta-analysis of composting technologies reveals much potential for mitigating nitrogen losses. *Sci. Total Environ.* **2020**, *704*, 135401. [CrossRef]
20. Qu, J.; Zhang, L.; Zhang, X.; Gao, L.; Tian, Y. Biochar combined with gypsum reduces both nitrogen and carbon losses during agricultural waste composting and enhances overall compost quality by regulating microbial activities and functions. *Bioresour. Technol.* **2020**, *314*, 123781. [CrossRef]
21. Abdel-Shafy, H.I.; Mansour, M.S.M. Solid waste issue: Sources, composition, disposal, recycling, and valorization. *Egypt. J. Pet.* **2018**, *27*, 1275–1290. [CrossRef]
22. Oladele, S.O.; Adetunji, A.T. Agro-residue biochar and N fertilizer addition mitigates CO2-C emission and stabilized soil organic carbon pools in a rain-fed agricultural cropland. *Int. Soil Water Conserv. Res.* **2020**. [CrossRef]
23. Petric, I.; Selimbašić, V. Development and validation of mathematical model for aerobic composting process. *Chem. Eng. J.* **2008**, *139*, 304–317. [CrossRef]
24. Mason, I.G. Mathematical modelling of the composting process: A review. *Waste Manag.* **2006**, *26*, 3–21. [CrossRef] [PubMed]
25. Wang, Y.; Niu, W.; Ai, P. Assessing thermal conductivity of composting reactor with attention on varying thermal resistance between compost and the inner surface. *Waste Manag.* **2016**, *58*, 144–151. [CrossRef] [PubMed]
26. Wang, Y.; Ai, P.; Cao, H.; Liu, Z. Prediction of moisture variation during composting process: A comparison of mathematical models. *Bioresour. Technol.* **2015**, *193*, 200–205. [CrossRef] [PubMed]
27. Ma, S.; Sun, X.; Fang, C.; He, X.; Han, L.; Huang, G. Exploring the mechanisms of decreased methane during pig manure and wheat straw aerobic composting covered with a semi-permeable membrane. *Waste Manag.* **2018**, *78*, 393–400. [CrossRef]
28. He, X.; Han, L.; Ge, J.; Huang, G. Modelling for reactor-style aerobic composting based on coupling theory of mass-heat-momentum transport and Contois equation. *Bioresour. Technol.* **2018**, *253*, 165–174. [CrossRef] [PubMed]
29. Zambra, C.E.; Moraga, N.O.; Escudey, M. Heat and mass transfer in unsaturated porous media: Moisture effects in compost piles self-heating. *Int. J. Heat Mass Transf.* **2011**, *54*, 2801–2810. [CrossRef]
30. Zambra, C.E.; Rosales, C.; Moraga, N.O.; Ragazzi, M. Self-heating in a bioreactor: Coupling of heat and mass transfer with turbulent convection. *Int. J. Heat Mass Transf.* **2011**, *54*, 5077–5086. [CrossRef]
31. Alavi, N.; Sarmadi, K.; Goudarzi, G.; Babaei, A.A.; Bakhshoodeh, R.; Paydary, P. Attenuation of tetracyclines during chicken manure and bagasse co-composting: Degradation, kinetics, and artificial neural network modeling. *J. Environ. Manag.* **2019**, *231*, 1203–1210. [CrossRef] [PubMed]
32. Fernandez-Mena, H.; MacDonald, G.K.; Pellerin, S.; Nesme, T. Co-benefits and Trade-Offs from Agro-Food System Redesign for Circularity: A Case Study With the FAN Agent-Based Model. *Front. Sustain. Food Syst.* **2020**, *4*, 41. [CrossRef]
33. Bonifacio, H.F.; Rotz, C.A.; Richard, T.L. A Process-Based Model for Cattle Manure Compost Windrows: Part 1. Model Description. *Trans. ASABE* **2017**, *60*, 877–892. [CrossRef]
34. Lauwers, J.; Appels, L.; Thompson, I.P.; Degrève, J.; van Impe, J.F.; Dewil, R. Mathematical modelling of anaerobic digestion of biomass and waste: Power and limitations. *Prog. Energy Combust. Sci.* **2013**, *39*, 383–402. [CrossRef]
35. Walling, E.; Trémier, A.; Vaneeckhaute, C. A review of mathematical models for composting. *Waste Manag.* **2020**, *113*, 379–394. [CrossRef]
36. Li, S.; Fang, Y.; Wu, X. A systematic review of lean construction in Mainland China. *J. Clean. Prod.* **2020**, *257*, 120581. [CrossRef]
37. Heyman, A.; Law, S.; Berghauser Pont, M. How is Location Measured in Housing Valuation? A Systematic Review of Accessibility Specifications in Hedonic Price Models. *Urban Sci.* **2019**, *3*, 3. [CrossRef]

38. Jia, F.; Peng, S.; Green, J.; Koh, L.; Chen, X. Soybean supply chain management and sustainability: A systematic literature review. *J. Clean. Prod.* **2020**, *255*, 120254. [CrossRef]
39. Wijewardhana, U.A.; Meyer, D.; Jayawardana, M. Statistical models for the persistence of threatened birds using citizen science data: A systematic review. *Glob. Ecol. Conserv.* **2020**, *21*, e00821. [CrossRef]
40. Harris, R.C.; Sumner, T.; Knight, G.M.; White, R.G. Systematic review of mathematical models exploring the epidemiological impact of future TB vaccines. *Hum. Vaccin. Immunother.* **2016**, *12*, 2813–2832. [CrossRef] [PubMed]
41. Downs, S.H.; Black, N. The feasibility of creating a checklist for the assessment of the methodological quality both of randomised and non-randomised studies of health care interventions. *J. Epidemiol. Community Health* **1998**, *52*, 377. [CrossRef]
42. Shiflet, A.B.; Shiflet, G.W. *Introduction to Computational Science: Modeling and Simulation for the Sciences*, 2nd ed.; Princeton University Press: Princeton, NJ, USA, 2014; ISBN 9781400850556.
43. Lin, Y.P.; Huang, G.H.; Lu, H.W.; He, L. Modeling of substrate degradation and oxygen consumption in waste composting processes. *Waste Manag.* **2008**, *28*, 1375–1385. [CrossRef]
44. Oudart, D.; Robin, P.; Paillat, J.M.; Paul, E. Modelling nitrogen and carbon interactions in composting of animal manure in naturally aerated piles. *Waste Manag.* **2015**, *46*, 588–598. [CrossRef]
45. Calisti, R.; Regni, L.; Proietti, P. Compost-recipe: A new calculation model and a novel software tool to make the composting mixture. *J. Clean. Prod.* **2020**, 122427. [CrossRef]
46. Douglas, P.; Tyrrel, S.F.; Kinnersley, R.P.; Whelan, M.; Longhurst, P.J.; Walsh, K.; Pollard, S.J.T.; Drew, G.H. Sensitivity of predicted bioaerosol exposure from open windrow composting facilities to ADMS dispersion model parameters. *J. Environ. Manag.* **2016**, *184*, 448–455. [CrossRef] [PubMed]
47. Oudart, D.; Paul, E.; Robin, P.; Paillat, J.M. Modeling organic matter stabilization during windrow composting of livestock effluents. *Environ. Technol.* **2012**, *33*, 2235–2243. [CrossRef] [PubMed]
48. Ge, J.; Huang, G.; Huang, J.; Zeng, J.; Han, L. Particle-Scale Modeling of Methane Emission during Pig Manure/Wheat Straw Aerobic Composting. *Environ. Sci. Technol.* **2016**, *50*, 4374–4383. [CrossRef] [PubMed]
49. Vasiliadou, I.A.; Muktadirul Bari Chowdhury, A.K.M.; Akratos, C.S.; Tekerlekopoulou, A.G.; Pavlou, S.; Vayenas, D.V. Mathematical modeling of olive mill waste composting process. *Waste Manag.* **2015**, *43*, 61–71. [CrossRef]
50. Villaseñor, J.; Rodríguez Mayor, L.; Rodríguez Romero, L.; Fernández, F.J. Simulation of carbon degradation in a rotary drum pilot scale composting process. *J. Environ. Manag.* **2012**, *108*, 1–7. [CrossRef] [PubMed]
51. Zhang, Y.; Lashermes, G.; Houot, S.; Doublet, J.; Steyer, J.P.; Zhu, Y.G.; Barriuso, E.; Garnier, P. Modelling of organic matter dynamics during the composting process. *Waste Manag.* **2012**, *32*, 19–30. [CrossRef] [PubMed]
52. Lashermes, G.; Zhang, Y.; Houot, S.; Steyer, J.P.; Patureau, D.; Barriuso, E.; Garnier, P. Simulation of Organic Matter and Pollutant Evolution during Composting: The COP-Compost Model. *J. Environ. Qual.* **2013**, *42*, 361–372. [CrossRef]
53. St Martin, C.C.G.; Bekele, I.; Eudoxie, G.D.; Bristol, D.; Brathwaite, R.A.I.; Campo, K.-R. Modelling response patterns of physico-chemical indicators during high-rate composting of green waste for suppression of Pythium ultimum. *Environ. Technol.* **2014**, *35*, 590–601. [CrossRef]
54. Li, W.; Wu, C.; Wang, K.; Meng, L.; Lv, L. Nitrogen loss reduction by adding sucrose and beet pulp in sewage sludge composting. *Int. Biodeterior. Biodegrad.* **2017**, *124*, 297–303. [CrossRef]
55. Huang, G.; Wang, X.; Han, L. Rapid estimation of nutrients in chicken manure during plant-field composting using physicochemical properties. *Bioresour. Technol.* **2011**, *102*, 1455–1461. [CrossRef]
56. Petric, I.; Mustafić, N. Dynamic modeling the composting process of the mixture of poultry manure and wheat straw. *J. Environ. Manag.* **2015**, *161*, 392–401. [CrossRef]
57. Li, Z.; Lu, H.; Ren, L.; He, L. Experimental and modeling approaches for food waste composting: A review. *Chemosphere* **2013**, *93*, 1247–1257. [CrossRef] [PubMed]
58. Kabbashi, N. Sewage sludge composting simulation as carbon/nitrogen concentration change. *J. Environ. Sci.* **2011**, *23*, 1925–1928. [CrossRef]
59. Boniecki, P.; Dach, J.; Pilarski, K.; Piekarska-Boniecka, H. Artificial neural networks for modeling ammonia emissions released from sewage sludge composting. *Atmos. Environ.* **2012**, *57*, 49–54. [CrossRef]
60. Haug, R.T. *The Practical Handbook of Compost Engineering*; Lewis Publishers: Boca Raton, FL, USA, 1993; ISBN 0873713737.
61. Wang, Y.; Huang, G.; Zhang, A.; Han, L.; Ge, J. Estimating thermal balance during composting of swine manure and wheat straw: A simulation method. *Int. J. Heat Mass Transf.* **2014**, *75*, 362–367. [CrossRef]
62. Hamelers, H.V.M. *A Mathematical Model for Composting Kinetics*; Rulkens, W.H., van Straten, G., Eds.; Wageningen University: Wageningen, The Netherlands, 2001; ISBN 90-5808-445-0.
63. Hamelers, H.V.M. Modeling composting kinetics: A review of approaches. *Rev. Environ. Sci. Biotechnol.* **2004**, *3*, 331–342. [CrossRef]
64. Bonifacio, H.F.; Rotz, C.A.; Richard, T.L. A Process-Based Model for Cattle Manure Compost Windrows: Part 2. Model Performance and Application. *Trans. ASABE* **2017**, *60*, 893–913. [CrossRef]
65. Sun, W.; Huang, G.H.; Zeng, G.; Qin, X.; Yu, H. Quantitative effects of composting state variables on C/N ratio through GA-aided multivariate analysis. *Sci. Total Environ.* **2011**, *409*, 1243–1254. [CrossRef]
66. Bayram, A.; Kankal, M.; Ozsahin, T.; Saka, F. Estimation of the carbon to nitrogen (C:N) ratio in compostable solid waste using artificial neural networks. *Fresenius Environ. Bull.* **2011**, *20*, 3250–3257.

67. Hosseinzadeh, A.; Baziar, M.; Alidadi, H.; Zhou, J.L.; Altaee, A.; Najafpoor, A.A.; Jafarpour, S. Application of artificial neural network and multiple linear regression in modeling nutrient recovery in vermicompost under different conditions. *Bioresour. Technol.* **2020**, *303*, 122926. [CrossRef]
68. Díaz, M.J.; Eugenio, M.E.; López, F.; García, J.C.; Yañez, R. Neural Models for Optimizing Lignocellulosic Residues Composting Process. *Waste Biomass Valor* **2012**, *3*, 319–331. [CrossRef]
69. Mancebo, U.; Hettiaratchi, J.P.A. Rapid assessment of methanotrophic capacity of compost-based materials considering the effects of air-filled porosity, water content and dissolved organic carbon. *Bioresour. Technol.* **2015**, *177*, 125–133. [CrossRef] [PubMed]
70. Varma, V.S.; Kalamdhad, A.S.; Kumar, B. Optimization of waste combinations during in-vessel composting of agricultural waste. *Waste Manag. Res.* **2017**, *35*, 101–109. [CrossRef]
71. Chen, H.; Sun, S.; Zhang, B. Forecasting N 2 O emission and nitrogen loss from swine manure composting based on BP neural network. *MATEC Web Conf.* **2019**, *277*, 1010. [CrossRef]
72. Abdallah, M.; Abu Talib, M.; Feroz, S.; Nasir, Q.; Abdalla, H.; Mahfood, B. Artificial intelligence applications in solid waste management: A systematic research review. *Waste Manag.* **2020**, *109*, 231–246. [CrossRef] [PubMed]
73. Solle, D.; Hitzmann, B.; Herwig, C.; Pereira Remelhe, M.; Ulonska, S.; Wuerth, L.; Prata, A.; Steckenreiter, T. Between the Poles of Data-Driven and Mechanistic Modeling for Process Operation. *Chem. Ing. Tech.* **2017**, *89*, 542–561. [CrossRef]
74. Roy, D.; Azaïs, A.; Benkaraache, S.; Drogui, P.; Tyagi, R.D. Composting leachate: Characterization, treatment, and future perspectives. *Rev. Environ. Sci. Biotechnol.* **2018**, *17*, 323–349. [CrossRef]
75. Fernandez-Mena, H.; Gaudou, B.; Pellerin, S.; MacDonald, G.K.; Nesme, T. Flows in Agro-food Networks (FAN): An agent-based model to simulate local agricultural material flows. *Agric. Syst.* **2020**, *180*, 102718. [CrossRef]
76. Rotz, C.A.; Corson, M.; Chianese, D.; Coiner, C. *The Integrated Farm System Model*; Pasture Systems and Watershed Management Research Unit Agricultural Research Service United States Department of Agriculture: Washington, DC, USA, 2012.

MDPI
St. Alban-Anlage 66
4052 Basel
Switzerland
www.mdpi.com

*Processes* Editorial Office
E-mail: processes@mdpi.com
www.mdpi.com/journal/processes

www.ingramcontent.com/pod-product-compliance
Lightning Source LLC
Chambersburg PA
CBHW062037220926
51850CB00010B/2577
* 9 7 8 3 7 2 5 8 0 1 6 5 7 *